John Freely was born in New York in 1926 and joined the US Navy at the age of 17, serving during the last two years of World War II. He has a PhD in physics from New York University and did postdoctoral studies in the history of science at Oxford. He is Professor of Physics at Bosphorus University in Istanbul, where he has taught physics and the history of science since 1960. He has also taught in New York, Boston, London and Athens. He has written more than 40 books, including works on the history of science and travel. His most recent books on the history of science are *Flame of Miletus: The Birth of Science in Ancient Greece* (2012) and *Light from the East: How the Science of Medieval Islam Helped to Shape the Western World* (2011). His recent books on history and travel include *The Grand Turk, Storm on Horseback, Children of Achilles, The Cyclades, The Ionian Islands* (all I.B.Tauris), *Crete, The Western Shores of Turkey, Strolling Through Athens, Strolling Through Venice* and the bestselling *Strolling Through Istanbul* (all Tauris Parke Paperbacks).

To my beloved Toots.

Figure 1 Nicolaus Copernicus, from the 1554 Paris edition of his biography by Pierre Gassendi, presumably based on the self-portrait mentioned by Stimmer.

Celestial Revolutionary

Copernicus, the Man and His Universe

JOHN FREELY

I.B. TAURIS
LONDON · NEW YORK

Published in 2014 by I.B.Tauris & Co Ltd
6 Salem Road, London W2 4BU
175 Fifth Avenue, New York NY 10010
www.ibtauris.com

Distributed in the United States and Canada
Exclusively by Palgrave Macmillan
175 Fifth Avenue, New York NY 10010

Copyright © 2014 John Freely

The right of John Freely to be identified as the author of this work has been asserted by him in accordance with the Copyright, Designs and Patents Act 1988.

All rights reserved. Except for brief quotations in a review, this book, or any part thereof, may not be reproduced, stored in or introduced into a retrieval system, or transmitted, in any form or by any means, electronic, mechanical, photocopying, recording or otherwise, without the prior written permission of the publisher.

Every attempt has been made to gain permission for the use of the images in this book. Any omissions will be rectified in future editions.

ISBN: 978 1 78076 350 7
eISBN: 978 0 85773 490 7

A full CIP record for this book is available from the British Library
A full CIP record is available from the Library of Congress

Library of Congress Catalog Card Number: available

Typeset by Newgen Publishers, Chennai
Printed and bound in Sweden by ScandBook AB

CONTENTS

List of Illustrations vi
Introduction vii

1 'This Remote Corner of the Earth' 1
2 A New Age 13
3 The Jagiellonian University of Krakow 37
4 Renaissance Italy 53
5 The Bishopric of Warmia 65
6 The Little Commentary 75
7 The Letter Against Werner 85
8 The Frauenburg Wenches 99
9 The First Disciple 115
10 The First Account 131
11 Preparing the Revolutions 147
12 The Revolutions of the Celestial Spheres 163
13 The Copernican Revolution 185
14 Debating the Copernican and Ptolemaic Models 205
15 The Newtonian Synthesis 225
Epilogue Searching for Copernicus 245

Source Notes 251
Bibliography 265
Index 275

ILLUSTRATIONS

1. Nicolaus Copernicus, from the 1554 Paris edition of his biography by Pierre Gassendi, presumably based on the self-portrait mentioned by Stimmer   ii

2. The apparent motion of the Sun through the constellations Aries and Taurus (above); the apparent motion of Mars through the constellations Aries and Taurus (below)   18

3. Epicycle model for explaining the apparent retrograde motion of the planets (above); Ptolemy's equant model (below)   25

4. Aristotle's geocentric theory, Peter Apian, *Cosmographica*, 1539 (above); the Copernican heliocentric theory, *De revolutionibus*, 1543 (below)   94

5. The precession of the equinoxes (above); Copernican lunar model (middle); Copernican model for the solar anomaly (below) of a superior planet (left) and an inferior planet (right)   177

6. The Tychonic system (above); Kepler's first two laws of planetary motion (below)   193

7. Galileo's observations of the Moon with the telescope, from *Siderius nuncius* (The Starry Messenger), 1610   211

INTRODUCTION

This is a biography of the Polish astronomer Nicolaus Copernicus (1473–1543), who at the dawn of the Renaissance proposed the revolutionary theory that the earth and the other planets were in orbit around the sun, breaking with the geocentric cosmology that had been the world view since antiquity.

The heliocentric theory, as it was called, was published in 1543, just before Copernicus' death. His book is entitled *De revolutionibus orbium coelestium libri VI* (Six Books Concerning the Revolutions of the Heavenly Spheres), a truly revolutionary work whose reverberations were felt far beyond the realm of astronomy. During the first century after its publication, the Copernican theory was accepted by only a few astronomers, most notably Kepler and Galileo, but their new sun-centred astronomy sparked the seventeenth-century Scientific Revolution that climaxed with the new world system of Isaac Newton, the beginning of modern science.

Despite its great importance *De revolutionibus* appeared in only two editions, the first in 1543 and the second in 1566, and it was not translated into English until 1952, the year after I began earning my living as a physicist. Though his *De revolutionibus* has been called 'the book that nobody read,' it changed our view of the universe forever, breaking the bounds of the finite geocentric cosmos of antiquity and opening the way to the infinite and expanding universe of the new millennium.

CHAPTER 1
'THIS REMOTE CORNER OF THE EARTH'

Sigismondo de' Conti, the papal secretary, noted in his chronicle during the spring of 1500 that 'All the world is in Rome.' A few days before Christmas 1499 Pope Alexander VI Borgia had declared that the following year would be a Jubilee, a period of special solemnity, in accordance with the decree published in 1471 by Pope Paul II which declared that each 25th year of the Christian era should be celebrated thusly. A special indulgence would be granted to all pilgrims who came to Rome and visited the four principal churches of the city, beginning with St Peter's, whose doors would be open night and day throughout the Jubilee. The celebrations went on throughout the year, and on Easter Sunday an estimated 200,000 pilgrims thronged St Peter's Square for the Pope's blessing. The pious monk Petrus Delphinus was led to exclaim 'God be praised, who has brought hither so many witnesses to the faith.'

Among the pilgrims was a young student named Nicolaus Copernicus, who had come from Poland to Italy in the autumn of 1496 to enrol in the faculty of law at the University of Bologna. The Italian Renaissance was in full bloom and Copernicus was in Rome at the height of its glory, before returning home the following year. He came back to Italy later that year to study medicine at the University of Padua for two years, before going to the University of Ferrara, where in 1503 he received a doctorate in law. He then returned to what he later called 'this remote corner of the earth,' in present-day northern Poland, where he would remain for the rest of his days.

One of the earliest biographies of Copernicus, a somewhat unreliable work in Latin published in 1654 by the French philosopher and astronomer Pierre Gassendi, gives his name as Nicolai Copernici, one of many forms that appear in various sources, including the astronomer's own correspondence, Nicolaus Copernicus being the one now generally used.

Copernicus was born on 19 February 1473 in a house on Saint Anne's Street in Thorn (Torun), a town on the Vistula, 110 miles south of Danzig (Gdansk) and 110 miles northwest of Warsaw, in what was then Royal Prussia, a region of the Kingdom of Poland. He was named after his father, Niklas Koppernigk, but afterwards followed the academic custom of the time and Latinized his name as Nicolaus Copernicus.

The Koppernigk family were originally German-speakers who migrated eastward to the province of Silesia in the thirteenth century, settling in the town known today as Koperniki, in present-day southeast Poland, close to the Czech border. Around 1350 the family moved to Krakow, capital of the Kingdom of Poland, where Niklas Koppernigk, the astronomer's great-great grandfather, was made a citizen in 1396. The astronomer's father, also named Niklas Koppernigk, first appears in records in 1448 as a prosperous merchant dealing in copper, which he sold mostly in Danzig, the Polish port city at the mouth of the Vistula. Around 1458 he moved from Krakow to Thorn, where a few years later he married Barbara, daughter of Lucas Watzenrode the Elder, a wealthy merchant and city councillor.

The Watzenrodes also originated from Silesia, having taken their name from their native village of Weizenrodau near Schweidnitz, which they left for Thorn after 1360. Lucas Watzenrode the Elder was born in Thorn in 1400 and in 1436 he married Katherina von Rüdiger. Katherina was a widow, having previously been married to Henrich Pechau, a town councillor of Thorn, by whom she had a son, Johann Peckau, who would be like another uncle to the young Nicolaus Copernicus.

Lucas Watzenrode the Elder died in 1462, leaving three children: Barbara, the astronomer's mother; Christina, who in 1459 married Tiedeman Van Allen, a prosperous merchant serving in the last quarter of the fifteenth century eight one-year terms as Mayor of Thorn; and Lucas Watzenrode the Younger, who would become Bishop of Warmia

(Ermeland), the region between Pomerania and Masuria in northeastern Poland, one of the four provinces into which the Duchy of Prussia was then divided, with the Estates of Royal Prussia to its west and the Kingdom of Poland to its south. The Watzenrodes were further related by marriage to wealthy burgher families of Krakow, Danzig and Thorn, as well as prominent noble families of Royal Prussia.

Thorn was founded on the site of an old Polish settlement by the Teutonic Knights, who built a castle there in 1230. Three years later the Grand Master of the Teutonic Knights, Hermann von Salza, together with his associate Hermann Balk, signed the foundation charters for Thorn and the nearby city of Kulm (Chelmno). These were among the seventy places or more in Prussia founded by the Teutonic Knights, each being protected by a castle and often endowed with a church. The sense of security given by these castellated settlements attracted the surrounding farmers, and soon developed into towns and cities with craftsmen and traders, each of the communities surrounded by defence walls and interconnected by roads.

The Teutonic Knights were one of three orders – the others being the Knights Templars and the Knights Hospitallers of St John – founded during The Crusades, their purpose being to aid Christian pilgrims to the Holy Land by building hospices and hospitals for them as well as fighting alongside the crusaders. Their emblem, a black cross on a white field, contrasted with the red cross on white of the Knights Templars and the white cross on red of the Knights Hospitallers.

The Order of the Teutonic Knights was founded at the end of the twelfth century at Acre in the Gulf of Haifa, which had been captured by the army of the Third Crusade in 1191 after a memorable siege. Following the defeat of the Christian forces in the Levant, the Order moved to Transylvania in 1211 to help defend Hungary against an invasion by a Turkic tribe known as the Cumans. Then in 1226 Duke Conrad I of Mazovia invited Hermann von Salza to move his knights into the Baltic region to conquer and Christianize the pagans known as the Old Prussians. Pope Honorius III had already called for a crusade against the Prussians, but this had been unsuccessful and Duke Conrad was thus led to bring in the Teutonic Knights, giving them a large grant of land in Culmerland, the region around Kulm, as well as any territory they might conquer, putting them only under the authority of the Holy See.

The Teutonic Knights slaughtered and enslaved the Prussians and seized their lands, and by the mid-fourteenth century they had taken control of most of the northern tier of what is now Poland. The Kingdom of Poland, much reduced in size because of the incursions of the Teutonic Knights and other powers, began to revive under Casimir III, the Great (r. 1333–70), the last king of the Piast dynasty, which had ruled since the end of the tenth century.

When Casimir began his reign, the Polish economy was ruined and the country depopulated and devastated by continual war. When he died he left a prosperous kingdom that had doubled in size, mostly through the addition of territory in what is today the Ukraine. He had reformed the institutions of the kingdom, sanctioned a code of laws, built many new castles, and, with the permission of Pope Urban V, founded a Studium Generale in Krakow, the first institution of higher learning in Poland. As part of his effort to repopulate the kingdom he encouraged Jews to resettle in Poland in large numbers, protecting them as 'people of the king.' As a result some 70 per cent of Ashkenazi Jews trace their origin to Poland in the time of Casimir the Great.

Casimir had no legal sons, and so he arranged for his sister Elisabeth, Dowager Queen of Hungary, and her son Louis, King of Hungary, to be his successors to the Polish throne. After his death 1370, Louis was proclaimed King of Poland, though his mother Elisabeth was the power behind the throne until her death in 1380. When Louis died in 1382, he was succeeded by his eldest surviving daughter, Mary, who became Queen of Hungary. But the Polish nobility were opposed to a personal union with Hungary, and they chose Mary's younger sister, Hedwig, who on 15 October 1384 was crowned in Krakow as King Jadwiga of Poland, not long after her tenth birthday. (Her official title was 'king' rather than 'queen', signifying that she was a sovereign in her own right and not just a royal consort.)

Two years later Jadwiga was betrothed to Jogaila, Grand Duke of Lithuania, an illiterate heathen who was about 24 at the time. Jogaila had agreed to adopt Christianity and promised to return to Poland lands that had been 'stolen' by its neighbours. Jadwiga had misgivings about the marriage, for she had heard that Jogaila was a filthy bear-like barbarian, cruel and uncivilized, and so she sent one of her knights, Zawisza the Red, to see if her proposed husband was really human. Zawisza reported that Jogaila was beardless, clean and civilized, and though an unlettered heathen he seemed to have a high regard for Christian culture. Therefore, Jadwiga went ahead with the marriage, which was held in Krakow Cathedral on 4 March 1386, two weeks after Jogaila was baptized. Immediately after the wedding Archbishop Bodzanta crowned Jogaila, who became King of Poland as Wladyslaw II Jagiello, beginning a reign that would last for 48 years. Thus started the illustrious Jagiellonian dynasty, which reigned until 1572; its dynasts ruling as Kings of Poland and Grand Dukes of Lithuania.

Wladyslaw and Jadwiga reigned as co-rulers, and though Jadwiga had little real power she was very active in the political and cultural life of Poland. She led two expeditions into Ruthenia in 1387, when she was only thirteen, and recovered territory that had been transferred to Hungary during her father's reign. Three years later she personally opened negotiations with the Teutonic Knights. Jadwiga gave birth to a daughter on 22 June 1399, but within a month both mother and child died.

Jadwiga was renowned for her charitable works and religious foundations, which led to her canonization as a saint in 1997 by the Polish Pope John Paul II. One of the legacies in her last will and testament provided for the restoration of Krakow's Studium Generale, otherwise known as Krakow Academy, a bequest that was faithfully carried out by King Wladyslaw, creating the institution known today as the Jagiellonian University of Krakow.

A Polish–Lithuanian army broke the power of the Teutonic Knights at the Battle of Tannenberg in 1410. This war ended with the First Peace of Thorn, signed in February 1411. According to the terms of this treaty, the Teutonic Knights held on to most of their territory through the control of their fortified cities and towns, though their subjects grew increasingly rebellious under the harsh rule of the Order.

During the next quarter of a century the Polish Crown fought the Teutonic Knights in a series of three wars that devastated Prussia, though with no territorial loss for the Order. In 1440 the gentry of Thorn joined with other towns to form the Prussian Confederation, which in 1454 rose up in revolt against the Teutonic Knights, beginning the Thirteen Years' War, in which they were aided by Casimir IV Jagiellon, King of Poland and Grand Duke of Lithuania. At the beginning of the revolt the people of Thorn stormed and captured the castle of the Knights and killed or imprisoned its defenders. The rebellion finally ended on 19 October 1466 with the Second Peace of Thorn. According to the terms of the treaty, the western part of the Order's territory along the lower Vistula came under Polish suzerainty as the Estates of Royal Prussia, which included Thorn and Danzig, while the wealthy see of Warmia became a separate dominion ruled by its bishop under the Polish Crown.

The Teutonic Knights retained only the hinterland of the port of Königsberg bounded on the southwest by Warmia. The Peace of Thorn was reaffirmed on 8 April 1525 by the Treaty of Krakow, which gave the Grand Master of the Teutonic Knights hereditary possession of the Order's territory, then known as 'Ducal Prussia', as a fief of the Polish Crown.

Such was the political chequerboard of the 'remote corner of the earth' where Copernicus was born and spent most of his life. His father had moved to Thorn during the Thirteen Years' War against the Teutonic Knights, and he lent money to the city to help support the soldiers of the Crown who were defending it as well as paying for a bridge across the Vistula, later serving as magistrate and alderman. Copernicus' maternal grandfather, Lucas Watzenrode the Elder, had fought against the Teutonic Knights in the Thirteen Years' War, in which he was wounded. He is listed in the *Thorner Bürger Buch*, the registry of the citizens of Thorn, as a landowner, businessman, judge and councilman, the type of burgher who had formed the core of the resistance to the Order of the Teutonic Knights.

Thorn was a member of the Hanseatic League, an alliance of trading cities and their guilds that held a trade monopoly along the northern tier of Europe from the Baltic to the North Sea. The commercial activities that led to the formation of this alliance originated in 1159 in the northern German port city of Lübeck, 'Queen of the Baltic', after it was rebuilt by Duke Henry the Lion of Saxony. Lübeck became a base for merchants in Saxony and Westphalia to trade farther afield along the coast from the North Sea and the Baltic and up rivers into the hinterland to cities like Thorn and Krakow, forming guilds known as *Hansa*, which bound the member cities to come to one another's aid with ships and armed men. The formal founding of the League came in 1356 at Lübeck, when representatives of the member cities met in the town hall and ratified the charter of the first *Hansetag*, or Hanseatic Diet.

'This Remote Corner of the Earth'

Lübeck and other cities of the League built trading posts called *kontor*, founding them as far afield as the inland Russian port of Novgorod, Bergen in Norway, and London. The London *kontor*, established in 1320, was west of London Bridge near Upper Thames Street on the present site of Cannon Street station. Like Hanseatic *kontors* elsewhere, the trading post in London developed into a walled community with its own warehouses, weigh house, offices, houses and church. Beside the *kontors*, each of the Hanseatic ports had a warehouse run by a representative of the League, those in England located in Bristol, Boston, Bishop's Lynn (now King's Lynn), Hull, Ipswich, Norwich, Yarmouth (now Great Yarmouth) and York. Krakow, Thorn and Danzig had Hansa representatives, the latter becoming the largest city in the League due to its control of Polish grain exports. By the beginning of the sixteenth century, Danzig had a population of more than 35,000, while Krakow, the capital of the Polish Kingdom, had about 20,000 inhabitants and Thorn some 10,000.

A fifteenth-century chronicler describes Thorn in the time of Copernicus: 'Thorn with its beautiful buildings and its roofs of gleaming tile is so magnificent that almost no town can match it for beauty of location and splendor of location.' The population of the city is now 20 times greater than it was in the fifteenth century, but the old walled town on the right bank of the Vistula is almost miraculously preserved, with its many Gothic buildings, all in brick, laid out along the medieval network of narrow streets and around the cobbled main square, still dominated by the Old Town Hall built in 1274 and extended in the late sixteenth century. When viewed from the Vistula the old town is still much the same as it appears in a lithograph done in 1684 by Christoph Hartknoch, lacking only the sailing barges that Copernicus would have seen in his youth, making their way along the river to and from the docks below the city walls.

Nicolaus Copernicus was the youngest of four children, the others being his brother Andreas and his sisters Barbara and Katherina. When he was seven years old, the family moved from Saint Anne's Street to a larger house on the main square of Thorn, where the city's weekly market was held. By that time he had started in the parochial school at the nearby Church of St Johann, whose renown attracted students from all over Poland. There his studies included mathematics and Latin, which was not only the universal academic language of Europe but was used in the liturgy at the Church of St Johann and spoken by the merchants of the Hanseatic League who traded in Thorn.

King Casimir IV visited Thorn in 1485, accompanied by his court, disembarking from the royal barge beside the main gate of the city, the entire populace there to greet him. Casimir spent six weeks in Thorn, dining in turn at the houses of the various notables, and so the young Nicolaus Copernicus would have met the king several times, for his extended family included the most influential people in the city.

Niklas Koppernigk died some time between 18 July 1483 and 19 August 1485, the former date marked in the last record of his financial affairs, and the latter by a reference to him as deceased. He was buried in the Church of St Johann, where his portrait can still be seen on his funerary monument; a tall, slim figure with a moustache and long black hair, shown on his knees with his hands joined in prayer. His son Nicolaus would have been among the mourners at the funeral, left without a father before he had even entered his teens.

Barbara Koppernigk never remarried, and she continued to live in the house on the market square with her children until she died, passing away some time between 1495 and 1507. Her oldest daughter, Barbara, left the house to become a nun at the Benedictine convent in Kulm. The youngest girl, Katherina, married a merchant from Krakow, Bartholomaeus Gertner, who had moved to Thorn and become a city councillor. The Gertners moved into the Koppernigk house, where their five children were born and they continued to live until at least 1507.

Nicolaus and his older brother Andreas were taken in hand after their father's death by their uncle Lucas Watzenrode, who looked after their education. Lucas had studied at the Jagiellonian University in Krakow in the years 1463–4, after which he went on to the University of Cologne, where he received an MA in 1468. He then completed his education at the University of Bologna, where in 1473 he was awarded a doctorate in canon law.

After receiving his doctorate Lucas returned to Thorn, where he found employment as a school teacher. At the school he became involved with the principal's daughter, described by a contemporary chronicler as a 'pious virgin'. The result of this affair was an illegitimate son named Philipp Teschner, who later became Mayor of Braunsberg

(Braniewo), a town in east Prussia, where he was a prominent supporter of the Protestant Reformation.

Lucas left the school before his bastard was born, giving up teaching and embarking on a career in the Church. The following year he was appointed Canon of Leczyca, a town southeast of Thorn. During the years 1477–88 he worked as a close collaborator with Sbigneus Olesnicki the Younger, nephew of Cardinal Sbigneus Olesnicki the Elder, the most powerful man in Poland after King Casimir IV. Lucas took up residence with Sbigneus the Younger at Gnesen, 60 miles southwest of Thorn. While he was there, he used his influential connections to secure new prebends, or stipends: first the canonry of Wladyslaw in 1478, then Warmia in 1479 and Gnesen in 1485. He was finally ordained as a priest in 1487.

The Second Peace of Thorn in 1466 had removed Warmia from the control of the Teutonic Knights and placed it under the sovereignty of the Polish Crown as part of the province of Royal Prussia, although with special privileges that gave it some degree of autonomy under its bishop. The following year the cathedral chapter of Warmia elected Nicolaus von Tüngen as bishop, going against the wishes of King Casimir IV. The new bishop allied himself with the Teutonic Knights and King Matthias Corvinus of Hungary. This led to a conflict known as the War of the Priests, which began in 1478 when the army of the Polish Crown invaded Warmia, putting the town of Braunsberg under siege. The town withstood the siege, and the war ended the following year with the Treaty of Piotrkow Trybunalski. According to the terms of the treaty, King Casimir recognized von Tüngen as bishop and accepted the right of the cathedral chapter of Warmia to elect future bishops, provided that they were accepted by the Polish king and swore loyalty to him.

On 31 January 1489 von Tüngen resigned because of ill health, and soon afterwards the cathedral chapter elected Lucas Watzenrode as Bishop of Warmia. The new bishop was mitred by Pope Innocent VIII, once again against the explicit wishes of King Casimir, who had wanted the bishopric for his son Frederic. Watzenrode prevailed, and when Casimir died in 1492 the independence of the bishopric of Warmia was confirmed by his son and successor John I Albert.

Bishop Lucas numbered among his close friends several humanist scholars who were leading figures in the Renaissance, most notably

Jan Dlugosz, Conradus Celtes and Filippo Buonaccorsi, all three of whom had graduated from or lectured at the Jagiellonian University of Krakow. The young Nicolaus Copernicus would have met them as well as other learned friends of his uncle, putting him in touch with the humanist movement of the Renaissance at an early age.

Jan Dlugosz (1415–80), a graduate of the Jagiellonian University, was Canon at Krakow and later Archbishop of Lemberg. He too was a protégé of Sbigneus Olesnicki the Elder and wrote a biography of the Cardinal. Dlugosz was tutor to the children of Casimir IV, three of whom, John I Albert (r. 1492–1501), Alexander (r. 1501–6) and Sigismund I (r. 1506–48), would succeed their father in turn as King of Poland. He was sent by Casimir on diplomatic missions to the Papacy and the court of the Holy Roman Emperor, and was involved in the King's negotiations with the Teutonic Knights during the Thirteen Years' War and at the peace negotiations afterwards. Dlugosz is best known for his *Annale seu cronicae incliti Regni Poloniae* (Annals or chronicles of the famous Kingdom of Poland) and *Historiae Polonicae librii XII* (Polish Histories, in 12 books). The first of these works covers events not only in Poland but elsewhere in Europe from 965 up until the author's death in 1480, in which he synthesizes historical information with legends and possibly fiction.

Conradus Celtes (1459–1508) was born in Germany under his original name Konrad Bickel, which he Latinized when he began his higher studies, first at the University of Cologne and then at the University of Heidelberg. After finishing university he gave humanist lectures, first in central Europe and then in Rome, Florence, Bologna and Venice. His first book was *Ars versificandi et carminum* (The art of writing verses and poems), published in 1486. When he returned to Germany, he was brought to the attention of Emperor Frederick II, who named him Poet Lauraeate, after which he was given a doctoral degree by the University of Nuremberg. After making a lecture tour of the Holy Roman Empire, he travelled to Krakow and joined the Jagiellonian University, lecturing on mathematics, astronomy and the natural sciences. In Krakow he collaborated with other poets in founding a learned society based on the Roman academies, the Sodalitas Litterarum Vistulna (Literary Society of the Vistula). Celtes founded other branches of this society in Hungary, Austria and Germany, where he was made a professor at the University of Heidelberg. In 1497 he was called to Vienna by Emperor Maximilian I, who appointed him Teacher of the Art of Poetry and Conversation, with imperial privileges. This was the first time such an honour had been bestowed. In Vienna he lectured on the works of classical Greek and Latin writers and in 1502 founded the Collegium Poetarum, a college for poets. He was appointed Head of the Imperial Library founded by Maximilian, and collected numerous Greek and Roman manuscripts, his most notable discovery being the *Tabula Peutingeriana*, or *Peutinger Table*, the only known surviving map of the Roman Empire, with annotated itineraries for the aid of travellers. Celtes was working on the publication of the *Peutinger Table* when he died of syphilis in Vienna on 4 February 1508. The disease was then known as '*morbus gallicus*', or the 'French disease', which he had apparently contracted while lecturing in Italy. His most enduring influence was in historical studies, for he was the first to teach the history of the known world as a whole.

Filippo Buonaccorsi (1437–96) was born in San Gimignano in Tuscany. He took the surname Callimachus after he moved to Rome in 1462 and became a member of the Roman Academy of Julius Pomponius Laetus. The paganist views and licentious lifestyle of the academicians led Pope Paul II to have them all arrested in 1467, but they pleaded for mercy and were soon released. Buonaccorsi and other members of the Academy took part in an unsuccessful attempt to assassinate the Pope in 1468, after which he fled to Poland. When the Pope's agents searched the Academy, they found homosexual verses written by Buonaccorsi to the Bishop of Segni, Lucio Fazini. The Pope's persecution of the academicians came to a sudden end when he died of a stroke on 26 July 1471, supposedly while being sodomized by a page boy.

When Buonaccorsi arrived in Poland he first found employment with Gregory of Sanok, Bishop of Lemberg. Later he was hired by King Casimir IV as tutor of the royal children, together with Jan Dlugosz. He was named royal secretary in 1474, subsequently serving as ambassador to the Sublime Porte in Istanbul and then acting as the King's representative in Venice. Buonaccorsi collaborated with Conradus Celtes in founding the Sodalitas Litterarum Vistulna in Krakow. He spent the rest of his days lecturing at the Jagiellonian University of Krakow, as well as writing poetry and prose in Latin. His best known works are biographies of King Wladyslaw III, Cardinal Sbigneus Olesnicki the Elder and Bishop Gregory of Sanok, all of whom had been his patrons.

The cathedral of the Warmia bishopric was at Frauenburg (Frombork), a port town about 100 miles east of Danzig. Not far to the east of Frauenburg was the smaller town of Braunsberg, where Philipp Teschner was appointed as mayor after his father Lucas Watzenrode became Bishop of Warmia. Lucas had always acknowledged his illegitimate son, and so it would seem that he had arranged Teschner's appointment as mayor.

The Bishop's palace was at Heilsberg (Lidzbark Warminski), 140 miles northeast of Thorn, to which Lucas returned as often as he could to visit his family and look after his nephews Andreas and Nicolaus. Lucas had decided that the two boys would follow in his footsteps, beginning as canons in his own cathedral chapter in Frauenberg, for with his powerful position and influential connections he could ease their way to the top of the Catholic hierarchy in Poland, particularly in the case of Nicolaus, for whom he seemed to have had great expectations.

When Nicolaus was 15 his uncle Lucas sent him to the cathedral school at Wloclawek, some 30 miles up the Vistula, where he would be prepared for his higher studies. Most of the teachers at the school were graduates of the University of Krakow, the most notable being Dr Nicolaus Wodka, who Latinized his name as Abstemius. Abstemius was a specialist in gnomonics, the study of shadows cast by a gnomon, the pointer on a sundial, and Nicolaus probably studied astronomy with him. There is a tradition that the sundial on the south side of Wloclawek Cathedral was constructed by Copernicus in collaboration with Abstemius.

After Nicolaus graduated from the school at Wloclawek, he and Andreas were sent by their uncle Lucas to his alma mater, the Jagiellonian University in Krakow. And so, after the arrangements had been made, Nicolaus and his brother set out from Thorn to Krakow in the autumn of 1491, beginning a journey that would eventually bring about an intellectual revolution and change a world view that had been held since antiquity.

CHAPTER 2
A NEW AGE

The latter half of the fifteenth century was the twilight of the medieval period and the dawn of a new era, the Renaissance of Western Europe. In 1453, 20 years before the birth of Copernicus, the Turks captured Constantinople, the capital of the Byzantine Empire and the Christian continuation of the Roman Empire. Two years later the Gutenberg Bible was printed. Nineteen years after Copernicus was born, Christopher Columbus discovered America, opening up a New World at the beginning of a new age.

The European Renaissance was the culmination of a 1,000 years of development that began after the collapse of Graeco-Roman civilization. The Roman Empire had been divided since AD 330, when Constantine the Great moved his capital to the city of Byzantium on the Bosphorus, renaming it Constantinople. The Western Empire finally came to an end in 480 with the death of Julius Nepos, the last Emperor of the West, who was assassinated in Dalmatia, the last remnant of his domain. Thenceforth, the Emperor in Constantinople was sole ruler of what remained of the Empire.

By the end of the fifth century AD the Roman Empire had been reduced to the predominately Greek-speaking East, where Christianity was rapidly supplanting the worship of the ancient Graeco-Roman deities. The heart of the Empire was now Asia Minor, where a Greek was more likely to be called a *Rhomaios*, or Roman, rather than a Hellene, which had come to mean a pagan, while the people of Constantinople referred to themselves as *Byzantini*, or Byzantine, and were Christian. Modern historians consider the end of the fifth century to be a watershed in the history of the Empire, which thenceforth is generally referred to as Byzantine rather than Roman.

The peak of the Byzantine Empire came under Justinian I (r. 527–65), who reconquered many of the lost dominions of the Empire, so that the Mediterranean once again became a Roman sea. Justinian I also broke the last direct link with the classical past when in 529 he issued an edict forbidding pagans to teach. As a result the ancient Platonic Academy in Athens was closed, ending an existence of more than nine centuries, as its teachers went into retirement or exile.

By the end of the eighth century, after successive invasions by the Persians, Arabs and Slavs, the Empire was reduced to little more than Asia Minor and a few enclaves in Greece, Italy and Sicily. Athens was utterly destroyed by the barbarian Heruli around 590, and was virtually uninhabited for centuries afterwards, while Alexandria was taken by the Arabs in 639, its great library having been destroyed two centuries earlier by a mob of Christian fanatics. The great centres of the ancient Graeco-Roman world had fallen, leaving only Constantinople as a beleaguered bastion of a Christianized remnant of classical civilization.

The Library of Alexandria had preserved the writings of all the Greek writers from Homer onward. After its destruction, all of the original works of Greek philosophy and science were lost, but copies of many of them survived and eventually made their way to Western Europe by a number of routes and through various chains of translation.

The first Greek philosophers of nature had emerged in the sixth century BC in the Greek colonies on the Aegean coast of Asia Minor and its offshore islands, as well as in Magna Graecia, the Greek cities in southern Italy and Sicily. They were known as *physikoi*, or physicists, from the Greek *physis*, meaning 'nature' in its widest sense, for they were the first who tried to explain phenomena on natural rather than supernatural grounds. Now called the Presocratics, they included Thales, Anaximander, Anaximenes, Pythagoras, Xenophanes, Heraclitus, Parmenides, Empedocles and Anaxagoras.

Anaxagoras, one of the last of the Presocratics, was born about 500 BC in Clazomenae, one of the Greek cities on the Aegean coast of Asia Minor, and when he was about 20 he moved to Athens, which emerged as the political and intellectual centre of the Hellenic world

A New Age

after the end of the Persian Wars in 479 BC, beginning the classical period in Greek history. Anaxagoras was the first philosopher to dwell in Athens, where he became the teacher of Pericles, who, in his famous funeral oration in 431 BC, honoured the Athenians who fell in the first year of the Peloponnesian War. He reminded his fellow citizens that they were fighting to defend a free and democratic society that was 'open to the world,' and whose 'love of the things of the mind' had made their city 'an education to Greece.'

Pericles was referring to the famous philosophical schools of Athens, the most renowned of which emerged in the following century: the Academy of Plato (427–347 BC), who had been a disciple of Socrates (469–399 BC), and the Lyceum of Aristotle (384–322 BC), who been Plato's pupil at the Academy.

Most of the great philosophers and scientists of the classical period taught in Athens, the most notable exceptions being Hippocrates of Kos (460–c.370 BC), the father of Medicine, and Democritus of Abdera (c.470–c.404 BC), who with his teacher Leucippus formed the atomic theory.

Plato believed that mathematics was a prerequisite for the dialectical process that would give future leaders the philosophical insight necessary for governing a state. The mathematical study included arithmetic, plane and solid geometry, harmonics and astronomy. Harmonics involved a study of the physics of sound as well as an analysis of the numerical relations supposedly developed by the Pythagoreans in their researches on music. This led the Pythagoreans to believe that the cosmos was designed according to harmonious principles, as is evident not only in music but in the eternally recurring motions of the heavenly bodies, the eternal 'harmony of the celestial spheres.' Astronomy was studied not only for its practical applications, but for what it revealed of 'the true numbers' and 'true motions' behind the apparent movements of the celestial bodies.

Plato considered that philosophers should approach the study of nature, particularly astronomy, as an exercise in geometry. Through this idealized geometrization of nature, relations that were as certain as those in geometry could therefore be obtained. As Socrates remarks in the *Republic*: 'Let's study astronomy by means of problems, as we do geometry, and leave the things in the sky alone.'

The main difficulty in Greek astronomy was to explain the apparent motion of the stars, the sun, the moon and the five visible planets. These are known as the celestial bodies and they all seem to rotate daily about a point in the heavens, namely the celestial pole, which is caused by the axial rotation of the earth in the opposite direction. Although the sun rises in the east and sets in the west each day, it appears to move back from west to east by about one degree from day to day, hence traversing the 12 signs of the zodiac in one year, which is a phenomenon caused by the orbiting of the earth around the sun.

The apparent path of the sun through the zodiac (i.e., the ecliptic) makes an angle of about 23.25 degrees with the celestial equator, which is the projection of the earth's equator out into space. This is explainable by the fact that the earth's axis is tilted by about 23.25 degrees with respect to the perpendicular of the ecliptic plane, which is an inclination that causes the recurring cycle of the seasons.

The planets all trace paths near the ecliptic, which seems to go from east to west during the night alongside the fixed stars, while from one night to the next they move slowly back from west to east around the zodiac. Each planet also exhibits an apparent periodic retrograde motion, which appears as a loop when plotted on the celestial sphere. This is owed to the fact that in orbiting the sun the earth passes the slower outer planets and is itself passed by the swifter inner ones, in both cases making it appear that the planet is moving backwards for a time among the stars.

According to Simplicius ($c.490$–$c.560$), Plato posed a problem for those studying the heavens: to demonstrate 'on what hypotheses the phenomena [i.e., the "appearances", in this case the apparent retrograde motions] concerning the planets could be accounted for by uniform and ordered circular motions.'

The first solution to the problem was provided by Eudoxus of Cnidus ($c.400$–$c.347$ BC), a younger contemporary of Plato at the Academy. Eudoxus was the greatest mathematician of the classical period, credited with some of the theorems that would later appear in the works of Euclid and Archimedes. He was also the leading astronomer of his era, and had made careful observations of the celestial bodies from his observatory at Cnidus, on the southwestern coast of Asia Minor. Eudoxus suggested that the path of the five planets was the result of the uniform motion of four connected spheres, all of which had the earth as their centre, but with their axes inclined to one another and rotating at different speeds. The planet is attached to the equator of the innermost sphere, and the outermost sphere moves with the fixed stars. The motions of the sun and the moon were accounted for by three spheres each, while a single sphere sufficed for the daily rotation of the fixed stars, making a total of 27 spheres for the cosmos. Eudoxus' model, known as the theory of homocentric spheres, was elaborated upon by Callipus of Cyzicus (fl. 370 BC), who added two more spheres for the sun and moon and one more for Mercury, Venus and Mars, to make a total of 34 spheres. The theory of homocentric spheres was subsequently adopted by Aristotle as the physical model for his geocentric cosmos, using 55 planetary spheres plus another for the fixed stars.

Aristotle's writings are encyclopaedic in scope, including works on logic, metaphysics, rhetoric, theology, politics, economics, literature, ethics, psychology, physics, mechanics, astronomy, meteorology, cosmology, biology, botany, natural history and zoology. The main outlines of Aristotle's theory of matter and his cosmology derive from earlier Greek thought, which distinguished between the imperfect and transitory terrestrial world below the sphere of the moon and the perfect and eternal celestial region above. He took from Thales, Anaximander and Anaximenes the notion that there was one fundamental substance in nature, and reconciled this with Empedocles' concept of the four terrestrial elements – earth, water, air and fire – to which he added the *aether* of Anaxagoras, the quintessential element, as the basic substance of the celestial region.

According to Aristotle, the fundamental terrestrial substance, which he called *prostyle*, is completely undifferentiated. When this matter takes on various qualities it becomes one of the four terrestrial elements, and through further developments it takes on the form of the things seen in the world. Aristotle would describe this as matter taking on form. Thus, matter is the raw material; form is the collection of all the qualities that give an object its distinctive character. These two aspects of existence – matter and form – are inseparable, and can only exist in conjunction with one another.

Aristotle's cosmology arranged the four elements in order of density, with the immobile spherical earth at the centre surrounded by

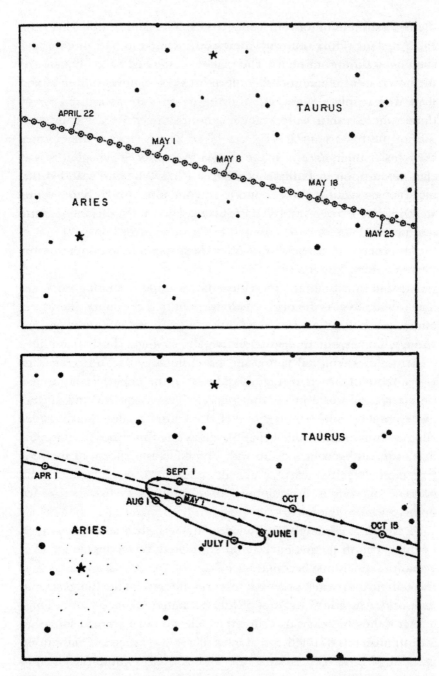

Figure 2 The apparent motion of the Sun through the constellations Aries and Taurus (above); the apparent motion of Mars through the constellations Aries and Taurus (below).

concentric shells of water (the ocean), air (the atmosphere) and fire, the latter including not only flames but extraterrestrial phenomena such as lightning, rainbows and comets. The natural motion of the four terrestrial elements moves towards their natural place, so that if earth is displaced upward in air and released, it will fall straight down, whereas air in water will rise, and so does fire in air. This linear motion of the terrestrial elements is temporary since it ceases when they reach their natural place. Aristotle's theory of motion argues that heavier objects fall faster than those that are light, and Aristotle also argued against the possibility of void. These two theories that we know are erroneous will dominate physics until the seventeenth century.

According to Aristotle, the celestial region begins at the moon, beyond which are the sun, the five planets and the fixed stars, all embedded in crystalline spheres rotating around the immobile earth. The celestial bodies are made of *aether*, whose natural motion is circular at constant velocity, so that the motions of the celestial bodies, unlike those of the terrestrial region, are unchanging and eternal.

Heraclides Ponticus (*c.*390–*c.*339 BC), named this way because he was a native of Heraclea on the Pontus (the Black Sea), was a contemporary of Aristotle, and had also studied at the Academy under Plato. His cosmology differed from that of Plato and Aristotle in at least two fundamental points, which may be due to the fact that after leaving the Academy he seems to have studied with the Pythagoreans. The first point of difference concerned the extent of the cosmos, which Heraclides thought to be infinite rather than finite. A second difference related to the apparent circling of the stars around the celestial pole, which, according to Heraclides, was actually due to the rotation of the earth on its axis in the opposite sense. Simplicius, in his commentary on Aristotle, writes that 'Heraclides supposed that earth is in the center and rotates while the heaven is at rest, and he thought by this supposition to save [i.e., account for] the phenomena.'

Aristotle's successor as director of the Lyceum was his son-in-law Theophrastus (*c.*371–287 bc), to whom he gave his enormous library, including copies of his complete works. Theophrastus headed the Lyceum for 37 years, during which time he reorganized and enlarged it, making him regarded as the second founder of the Lyceum.

More than 200 works, which are now mostly lost, are attributed to Theophrastus, whose interests was as encyclopedic as that of Aristotle. Theophrastus is known as the father of Botany thanks to his two extant works: *History of Plants* and *Causes of Plants*. His treatise *On Stones* marks the beginning of geology and mineralogy. His work on human behaviour, entitled *Characters*, is a witty description of the types of people living in Athens during his time, all of whom still seem to be represented in the modern city.

Two other schools of philosophy were founded in Athens late in the fourth century BC. These were not formal institutions like the Academy and the Lyceum, but more loosely organized groups gathered to discuss philosophy. One of the schools, known as the Garden, was founded by Epicurus of Samos (341–270 BC) and the other, the Porch, was established by Zeno of Citium (c.335–263 BC). The name of the first school came from the fact that Epicurus lectured in the garden of his house, while the second was named for the Stoa Poikile, or Painted Porch, in the Agora, the meeting place of Zeno and his disciples, the Stoics. Both Epicurus and Zeno created comprehensive philosophical systems that were divided into three parts – ethics, physics and logic – in which the last two were subordinate to the first, whose goal was to secure happiness.

The physics of Epicurus was based on the atomic theory, to which he added the new concept that an atom moving through the void could at any instant 'swerve' from its path. This eliminated the absolute determinism that had made the original atomic theory of Leucippus and Democritus unacceptable to those who, like the Epicureans, believed in free will. Zeno and his followers rejected the atom and the void, for they looked at nature as a continuum in all of its aspects – space, time and matter – as well as in the propagation and sequence of physical phenomena. These two opposing schools of thought about the nature of the cosmos – the Epicurean atoms in a void versus the continuum of the Stoics – have competed with one another from antiquity to the present, for they seem to represent antithetical ways of looking at physical reality.

After the death of Alexander the Great in 323 BC, the beginning of the Hellenistic period, the intellectual centre of the Greek

A New Age

world shifted from Athens to Alexandria, the new city that he had founded on the Canopic branch of the Nile. Alexandria became the capital of a powerful kingdom ruled by Ptolemy I (r. 305–280 BC), founder of the Ptolemaic dynasty that ruled Egypt for nearly three centuries.

The emergence of Alexandria as a cultural centre was largely due to the establishment of a school of higher studies known as the Museum, called thusly because it was dedicated to the Muses. The Museum and its famous Library were founded by Ptolemy I and further developed by his son and successor Ptolemy II (r. 283–225 BC). By law the Library was required to obtain a copy of every work written in the Greek world, and by the time of Ptolemy III (r. 247–221 BC) it was reputed to have a collection of more than half a million parchment rolls, including everything written from Homer onward.

The first scientist to head the Library was Eratosthenes of Cyrene (*c*.275–*c*.195 BC), and to draw a map of the known world on a system of meridians of longitude and parallels of latitude, which, together with observations with a gnomon, allowed him to make an accurate estimation of the earth's circumference.

Eratosthenes was a friend of Archimedes (*c*.287–212 BC), who dedicated to him the famous treatise *On Method*. Archimedes, who was from Syracuse in Sicily, probably studied at Alexandria under the pupils of Euclid (fl. *c*.295 BC), whose great work on geometry, the *Elements*, he quoted from extensively.

One of Archimedes' works, *The Sand-Reckoner*, mentions a revolutionary astronomical theory proposed by his older contemporary Aristarchus of Samos (*c*.310–287 BC), writing that

> Aristarchus of Samos has, however, enunciated certain hypotheses in which it results from the premises that the universe is much greater than that just mentioned. As a matter of fact, he supposes that the fixed stars and the sun do not move, but that the earth revolves in the circumference of a circle about the sun, which lies in the middle of the orbit, and that the sphere of the fixed stars, situated about the same center as the sun, is so great that the circle in which the earth is supposed to revolve has the same ratio to the distance of the fixed stars as the center of the sphere to its surface.

The last sentence is of particular significance, for it explains why there is no stellar parallax, or apparent displacement of the stars, when the earth moves in orbit around the sun in the heliocentric theory of Aristarchus. Even the nearest stars are so far away, compared to the radius of the earth's orbit around the sun, that their parallax is far too small to be detected by the naked eye. This effect was not observed until the mid-nineteenth century, when telescopes of sufficient resolving power had been developed.

The treatise in which Aristarchus presents his heliocentric theory has not survived, undoubtedly because the idea conflicted with the accepted belief that the earth was the stationary centre of the cosmos. He also seems to have believed that the earth was not only orbiting around the sun but also rotating on its own axis. Cleanthes of Assos, his contemporary, is quoted by Plutarch as holding that Aristarchus should be charged with impiety 'on the ground that he was disturbing the hearth of the universe because he sought to save [the] phenomena by supposing that the heaven is at rest while the earth is revolving along the ecliptic and at the same time is rotating about its own axis.'

The only work of Aristarchus that has survived is his treatise *On the sizes and distances of the Sun and the Moon*. Here the stellar and lunar sizes and distances were calculated from geometrical demonstrations based on three astronomical observations, together with an estimation of the earth's diameter. The results of these measurements led Aristarchus to conclude that the sun is about 19 times farther from the earth than the moon, and that the sun is approximately 6¾ times as large and the moon about 1/3 as large as the earth. All of his values are grossly underestimated, because of the crudeness of his observations, but his geometrical methods were sound. His finding that the sun is larger than the earth may have been what led him to propose his heliocentric theory.

The other Hellenistic mathematician comparable to Euclid and Archimedes is Apollonius of Perge, who flourished in Alexandria during the reigns of Ptolemy III and Ptolemy IV (r. 221–203 BC), as well as in Pergamum during the reign of Attalos I (r. 241–197 BC). His only surviving work is his treatise *On Conics*, the first comprehensive and systematic analysis of the three types of conic sections: the ellipse (of which the circle is a special case), parabola and hyperbola.

A New Age

Apollonius is also credited with formulating mathematical theories to explain the apparent retrograde motion of the planets. One of his theories has the planet moving round the circumference of a circle, known as the epicycle, whose centre itself moves round the circumference of another circle, called the deferent, centred at the earth. Another has the planet moving round the circumference of an eccentric circle, whose centre does not coincide with the earth. He also showed that the epicycle and eccentric circle theories are equivalent, so that either model can be used to describe retrograde planetary motion.

Aside from the great theoreticians of the Hellenistic era, there were also three gifted inventors. The first of these was Ctesibus of Alexandria (fl. *c.*270 BC), whose written works have been lost, but whose ideas and inventions have been preserved in the writings of his successor, Philo of Byzantium (fl. 250 BC) and Hero of Alexandria (fl. AD 62). Hero is famous for his steam engine, one of the *thaumata*, or 'miracle-working' devices described in his treatises *On Automata* and *Pneumatic*. The first chapters of the latter work, which derive largely from Philo, describe experiments demonstrating that it is possible to produce a partial volume, contrary to Aristotelian doctrine. He also wrote a treatise on the reflection of light, the *Catoptrica*, which would play an important part in the development of early European studies of optics.

Hipparchus of Nicaea (fl. 147–127 BC) was the greatest observational astronomer of antiquity, whose observations were later used by Claudius Ptolemaeus (*c.* AD 100–*c.*170). All of his writings have been lost except for his first work, a commentary on the *Phainomena* of Aratus of Soli (*c.*310–*c.*240 BC), a Greek poem describing the constellations. *Phainomena* served to popularize the names of stars and constellations, which have been perpetuated in the modern world. It contains a catalogue of some 850 stars, for each of which Hipparchus gives the celestial coordinates, including those of a 'nova' or 'new star,' which suddenly appeared in 134 BC in the constellation Scorpio. He also estimated the brightness of the stars, assigning to each of them a 'magnitude,' which equalled one for the brightest stars and six for the faintest, a system still used in modern astronomy.

Hipparchus is famous for his discovery of the precession of the equinoxes, which, in brief terms, is the slow continuous shift in the orientation of the earth's rotational axis, whose direction remains perpendicular to the ecliptic. Hipparchus discovered this phenomenon by comparing his star catalogue with observations made 128 years earlier by the astronomer Timocharis, which led him to conclude that the annual precession moved at a rate of 45.2 arc seconds.

Hipparchus is also celebrated as a mathematician particularly for his development of spherical trigonometry, which he applied to problems in astronomy.

A younger contemporary of Hipparchus, the Greek astronomer and mathematician Theodosius of Bithynia ($c.160$-$c.100$ bc) is known for his *Sphaerica*, a treatise on the application of spherical geometry to astronomy, which was translated into Arabic and later into Latin. His work remained in use up until the seventeenth century.

Ancient Greek mathematical astronomy reached with the work of Claudius Ptolemaeus, known more simply as Ptolemy, who flourished in the mid-second century AD in Alexandria. The most influential of his writings is *Mathematical Synthesis*, a comprehensive work on theoretical astronomy better known by its Arabic name, the *Almagest*.

The topics in the *Almagest* are treated in logical order through the 13 books. Book I begins with a general discussion of astronomy, including Ptolemy's view that the earth is stationary 'in the middle of the heavens.' The rest of Book I and all of Book II are devoted principally to the development of the spherical trigonometry necessary for the whole work. Book III deals with the motion of the sun and Book IV with lunar motion, which is continued at a more advanced level in Book V along with solar and lunar parallax. Book VI is on eclipses; Books VII and VIII are on the fixed stars; and Books IX through XIII are on the planets.

Ptolemy's trigonometry and catalogue of stars are based on the work of Hipparchus, and his theory of epicycles and eccentrics is derived from Apollonius. The principal modification made by Ptolemy is that the centre of the epicycle moves uniformly with respect to a point called the equant, which is displaced from the centre of the deferent circle, a device that was to be the subject of controversy in later times.

The extant astronomical writings of Ptolemy also include the *Handy Tables, Planetary Hypotheses, Phases of the Fixed Stars, Analemma* and *Planisphaerium*; as well as a work on astrology called the *Tetrabiblos*; and

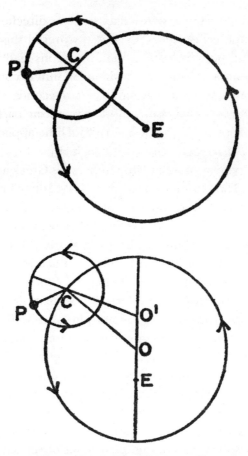

Figure 3 Epicycle model for explaining the apparent retrograde motion of the planets (above); Ptolemy's equant model (below).

treatises entitled *Optics, Geography* and *Harmonica*, the latter devoted to musical theory.

Galen (130–*c*.204), known in medieval Europe as the 'Prince of Physicians', was born in Pergamum in Asia Minor. He served his medical apprenticeship at the healing temple of Asclepios at Pergamum, where his work treating wounded gladiators gave him an unrivalled knowledge of human anatomy, physiology and neurology. He then travelled extensively and studied in many cities, including Smyrna, Corinth and Alexandria among others. In 161 he settled in Rome, where he was to spend most of the rest of his life serving as physician to three emperors. Galen's writings, translated into Arabic and Latin, served as the foundation of human anatomy and physiology up until the seventeenth century.

Another figure who came to study in Alexandria was Dioscorides Pedanius (fl. 50–70), from Anazarbus in southeastern Asia Minor, who later became a physician in the Roman army during the reigns of Claudius (r. 41–54) and Nero (r. 41–68). Dioscorides is regarded as the founder of pharmacology, renowned for his *De Materia Medica*, a systematic description of some 600 medicinal plants and nearly 1,000 drugs.

The last great mathematician of antiquity was Diophantus of Alexandria (fl. *c.*AD 250), who did for algebra and number theory what Euclid had done for geometry. The work of Diophantus is still a part of modern mathematics, studied under the heading of Diophantine Analysis.

Theon of Alexandria (*c.*335–*c.*405), who flourished in the second half of the fourth century, is noted in the Museum and Library, particularly for a passage in his commentary on a work by Ptolemy, where he states that 'certain ancient astrologers' believed that the points of the spring and autumn equinox oscillate back and forth along the ecliptic, moving through an angle of 8 degrees over a period of 640 years. This erroneous notion was revived in the so-called trepidation theory of Arab astronomers, and it survived in various forms up into the sixteenth century, when it was discussed by Copernicus.

After Justinian closed the Platonic Academy in Athens in 529, seven of its scholars went into exile, and in 531 they were given refuge by the Persian king Chosroes I, who appointed them to the faculty of the medical school at Jundishapur. Among them were Simplicius of Cilicia, Damascius and Isidorus of Miletus, the latter two having been the last directors of the Platonic Academy in Athens. The following year they were all allowed to return from exile, with Isidorus taking up residence in Constantinople and the others going back to Athens.

Justinian appointed Isidorus and his colleague Anthemius of Tralles to build the great church of Haghia Sophia in Constantinople, which was completed in 537 and survives virtually intact today, the supreme masterpiece of Byzantine architecture. Isidorus and Anthemius were the last mathematical physicists of antiquity, both of them authors of commentaries that are important links in the survival of Archimedes' writings.

Simplicius is known for his commentaries on Aristotle, whose ideas he defended against the criticism of his contemporary John Philoponus, particularly on the question of why a projectile, such as an arrow, continues moving after it is propelled. One of the last directors of the Platonic school in Alexandria, Philoponus (490–570) rejected the Aristotelian projectile theory presented by Simplicius, which was that the air displaced by the arrow flows back to push it from behind, a spurious effect called *antiperistasis*. Instead, Philoponus maintained that the arrow, when fired, receives an 'incorporeal motive force', an important concept that was revived in medieval Europe as the 'impetus theory'. Thus in the last days of antiquity an important debate took place on a fundamental scientific concept, just as the last light of classical Graeco-Roman civilization was being occulted at the onset of the Dark Ages.

Most of the Greek philosophical and scientific works that survived made their way through the Hellenized Syriac-speaking Christians of Mesopotamia to the Islamic world after 762, when the Abbasid Caliph al-Mansur founded Baghdad as his new capital. Baghdad emerged as a great cultural centre under al-Mansur (r. 754–75) and three generations of his successors, particularly al-Mahdi (r. 775–85), Harun al-Rashid (r. 786–809) and al-Ma'mun (r. 813–33). According to the historian al-Mas'udi (d. 956), al-Mansur 'was the first caliph to have books translated from foreign languages into Arabic,' including 'books by Aristotle on logic and other subjects, and other ancient books from classical Greek, Byzantine Greek, Pahlavi, Neopersian, and Syriac.' These translations were done at the famous *Bayt al-Hikma*, or House of Wisdom, a library established in Baghdad early in the Abbasid period.

The program of translation continued until the mid-eleventh century, both in the East and in Muslim Spain. By that time most of the important works of Greek science and philosophy were available in Arabic translations, along with commentaries on these works and the original treatises by Islamic scientists that had been produced in the interim. Thus, through their contact with surrounding cultures, scholars writing in Arabic were in a position to take the lead in science and philosophy, absorbing what they had learned from the Greeks and adding to it, hence beginning an Islamic renaissance, whose fruits were eventually passed on to Western Europe.

Some fragments of classical learning were also preserved in Latin Europe. These included writers of the Roman era, most notably Lucretius (*c*.94–50 BC), who wrote a superb didactic poem entitled *De Rerum Natura* (On the Nature of Things), based on the atomic theory of Democritus. *De Rerum Natura* became popular in medieval Europe, eventually leading to the revival of the atomic theory in the seventeenth century.

During the early medieval period, the attitude of Christian scholars was that the study of science was not necessary, for in order to save one's soul it is enough to believe in God, as St Augustine of Hippo (354–430) wrote in his *Enchiridion*: 'It is enough for Christians to believe that the only cause of created things, whether heavenly or earthly, whether visible or invisible, is the goodness of the Creator, the one true God, and that nothing exists but Himself that does not derive its existence from Him.'

The most important figure in the transmission of ancient Greek knowledge to early medieval Europe is Anicius Manlius Severinus Boethius (*c*.480–525). Boethius, who was from an aristocratic Roman family, held high office under the Ostrogoth king Theodoric, who had him imprisoned and executed. His best known work is his *Consolation of Philosophy*, written while he was in prison before his execution. The other works of Boethius fall into two categories: his translations from Greek into Latin of Aristotle's logical works, and his own writings on logic, theology, music, geometry and arithmetic. His writings were influential in the transmission to medieval Europe of the basic parts of Aristotle's logic and of elementary arithmetic.

The grand plan envisioned by Boethius was to transmit, in Latin, the intellectual achievements of the ancient Greeks. As he wrote in one of his commentaries: 'I shall transmit and comment upon as many works by Aristotle and Plato as I can get hold of, and I will try to show that their philosophies agree.' He also noted, in the introduction to his *Arithmetic*, a handbook for each of the disciplines of the *quadrivium*: arithmetic, geometry, astronomy and musical theory. The extant writings of Boethius also include translations and/or commentaries on at least five logical works by Aristotle, as well as translations of Euclid's *Elements*.

The revival of learning in Western Europe began in an Irish monastery in the sixth century. The first Irish monasteries were probably founded by priests from Britain and Gaul fleeing the Anglo-Saxon and Germanic invasions of the fifth and sixth centuries, as well as a few from the Greek world who came to both Ireland and Britain, bringing with them their language and certain texts that were not found elsewhere in Europe. St Columba (521–97) founded monasteries in Ireland, at Derry and Durrow, and then in 561 he went off into exile in Scotland along with 12 companions. He was granted land off the west coast of Scotland on the island of Iona, which became the centre of his evangelizing mission to the Picts.

St Benedict Biscop (*c*.628–90) was the founding abbot of the monasteries at Wearmouth and Jarrow on Tyne in Northumbria, making five trips to Rome to stock their libraries, which became famous throughout medieval Europe. On his third trip to Rome Benedict returned with Theodore of Tarsus (602–90), a Greek from Asia Minor who had studied in Antioch and Constantinople before joining a monastery in Rome, where in 668 Pope Vitalianus (r. 757–68) appointed him as Archbishop of Canterbury. Benedict also brought back Hadrian the African, a Greek-speaking Berber monk from North Africa who had twice turned down appointments to the archbishopric of Canterbury.

Theodore and Hadrian founded a monastic school at Canterbury that began what has been called the golden age of Anglo-Saxon scholarship. Theodore and Hadrian taught Greek and Latin as well as arithmetic and physical science. Students from the Wearmouth and Jarrow went on to become abbots of Benedictine monasteries elsewhere in England, spreading the knowledge of they had gained in Canterbury.

The Venerable Bede (674–735), writing in 731, says that at the age of seven he was entrusted to the care of Benedict Biscop at Wearmouth, where he began the studies that he would continue for the rest of his life there and at Yarrow. Bede's best known work is the *Historia ecclestiastica gentis Anglorus* (The Ecclesiastical History of the English People), completed in about 731. Bede wrote several *opera didascalica*, textbooks designed for vocational courses in monastic schools such as *notae* (scribal work), grammar (literary science) and *computus* (the art and science of telling time). His scientific writings include two treatises on astronomical timekeeping and a book called *De Rerum Natura*. The latter work presents his views on cosmology, which was based on the Aristotelian model in which the terrestrial region of the central earth is surrounded by the nested spheres of the celestial region.

Bede's tradition of learning was perpetuated by his student and disciple Egbert, Archbishop of York. The cathedral school that Egbert founded at York rivalled the famous monastic schools at Wearmouth and Jarrow, renowned not only in religious studies but in the seven liberal arts of the trivium and quadrivium as well as in literature and science, its library reputed to be the finest in Britain.

The most distinguished student at the York school in its early days was Alcuin of York (*c.*735–804), who had been enrolled there as a young boy by Egbert. On his way home from a mission to Rome in 781 Alcuin met Charlemagne, 'King of the Franks', who had in 768 inherited a realm now known as the Carolingian Empire, which including portions of what is now Germany and most of present-day France, Belgium and Holland, to which he eventually added more German territory, all of Switzerland, part of Austria, and half of Italy. Charlemagne asked Alcuin to join his court at Aachen as his principal advisor on ecclesiastical and educational matters.

Alcuin arrived in Aachen in 782 and was appointed Head of the Palace School, which had been founded by the earlier Merovingian kings to educate the royal children, principally in courtly ways and manners. Charlemagne wanted to broaden the education of his children and other members of his court to include the study of religion and the seven liberal arts. His students included Charlemagne himself as well as his sons Pepin and Louis, as well as young men sent to the court for their education along with clerics at the palace chapel.

Alcuin drew up a standardized curriculum and wrote textbooks and manuals for the Palace School, establishing the ancient *trivium* and *quadrivium* as the basis for education. He also introduced the various disciplines of the *computus*, including sufficient mathematics and astronomy for understanding the calendar, particularly the dating of Easter.

A New Age

Charlemagne issued an edict establishing monastic and cathedral schools throughout his realm. Students at the monastic and cathedral schools included not only clerics but also externs, or laymen, which led to an increase in Latin literacy among Charlemagne's subjects. Alcuin supervised the organization of these schools as well as their curricula and textbooks, which were based on the seven liberal arts. Many of his pupils at the Palace School went on to become bishops and abbots, significantly raising the educational level of the clergy, particularly those who directed and taught at the monastic and cathedral schools.

The Carolingian educational reforms produced the first major figure in what would become the new European science. This was Gerbert d'Aurillac (*c*.945–1003), who became Pope Sylvester II (r. 999–1003). Gerbert was born in humble circumstances in or near Aurillac, in southwestern France, and received his early education at the local Benedictine monastery of St Géraud. His precocious brilliance brought him to the attention of Count Borel of Barcelona who took him from the monastery in 967 as his protégé. He continued his studies in Barcelona under the aegis of Borel, who put him under the tutelage of Atto, Bishop of Vic. Gerbert concentrated on mathematics, probably studying the works of late Roman writers such as Boethius and Cassiodorus, which would have included some arithmetic, geometry (though without proofs) and astronomy, as well as Pythagorean number theory. He would also have learned the mensuration rules of ancient Roman surveyors.

While Gerbert was in Barcelona he seems to have come in contact with Islamic manuscripts, though probably only in Latin translation. Gerbert is credited with a treatise on the astrolabe entitled *De astrolabia*, as well as the first part of a work entitled *De utilitatibus astrolabi*, both of which show Arabic influence. The astrolabe, an astronomical instrument and calculating device, was invented by the ancient Greeks, probably by Hipparchus, and was widely used by Islamic astronomers.

Gerbert's attested writings include works on mathematics, one of which is a treatise on the abacus, a calculating device that is believed to have come to the Islamic world from China. He also seems to have used the Hindu-Arabic numbers which were subsequently adapted and became the basis for the ones used today. Thus it would appear that Gerbert was one of the first European scholars to make use of the Graeco-Islamic scientific heritage in developing the new science that was beginning to emerge in Western Europe.

Gerbert came to the attention of Otto I (r. 962–73), the Holy Roman Emperor, who was then residing in Rome. Otto arranged for Gerbert's assignment to Adalboro, Archbishop of Rheims, who was appointed Master of the Cathedral School there, which he reorganized with such success that students flocked to it from all parts of the Empire. Gerbert's students are known to have gone on to teach at eight other cathedral schools in northern Europe, where they emphasized the mathematics and astronomy he had learned from Islamic sources in Spain.

The development of European science began to accelerate when the heritage of Graeco-Islamic learning became available to Latin scholars in the West, beginning with Gerbert d'Aurillac.

Adelard of Bath (fl. 1116–42) was the leading figure in the European acquisition of Arabic science. In the introduction to his *Questiones Naturales*, addressed to his nephew, Adelard writes of his 'long period of study abroad,' first in France, where he studied at Tours and taught at Laon. He then went on to Salerno, Sicily, Tarsus, Antioch and probably also to Spain, spending a total of seven years abroad.

Adelard may have learned Arabic in Spain, for his translation of the *Astronomical Tables* of the Arabic scholar al-Khwarizmi (fl. *c.*828) was from the revised version of the Andalusian astronomer Maslama al-Majriti (d. 1034). The *Tables*, comprising 37 introductory chapters and 116 listings of celestial data, provided Christian Europe with its first knowledge of Graeco-Arabic-Indian astronomy and mathematics, including the first tables of the trigonometric sine function to appear in Latin. He also translated the *Introduction to Astrology* of Abu Ma'shar (787–886).

Adelard may also have been the author of the first Latin translation of another work by al-Khwarizmi, *De Numero Indorum* (Concerning the Hindu Art of Reckoning). This work, probably based on an Arabic translation of works by the Indian mathematician Brahmagupta (fl. 628), describes the Hindu numerals that eventually became the digits used in the modern Western world. The new notation came to be known as that of al-Khwarizmi, corrupted to 'algorism' or 'algorithm', which now means a procedure for solving a mathematical problem in a finite number of steps that often involves repetition of an operation.

Adelard was probably also the first to give a full Latin translation of Euclid's *Elements*, in three versions. The second of these became very popular, beginning the process that led to Euclid's domination of medieval European mathematics.

Adelard says that his *Questiones Naturalis* was written to explain 'something new from my Arab studies.' The *Questiones* are 76 in number, 1–6 dealing with plants, 7–14 with birds, 15–16 with mankind in general, 17–32 with psychology, 33–47 with the human body, and 48–76 with meteorology and astronomy. Throughout he looks for natural rather than supernatural causes of phenomena, a practice that would be followed by later European writers. The *Questiones Naturalis* remained popular throughout the rest of the Middle Ages, with three editions appearing before 1500, as well as a Hebrew version. Adelard also wrote works ranging from trigonometry to astrology and from Platonic philosophy to falconry. His last work was a treatise on the astrolabe, in which once more he explained 'the opinions of the Arabs,' this time concerning astronomy.

Toledo became a centre for translation from the Arabic after its recapture from the Moors in 1085 by Alfonso VI, King of Castile and León, the first major triumph of the *reconquista*, the Christian reconquest of al-Andalus. Gerard of Cremona (1114–87), the most prolific of all the Latin translators from the Arabic, first came to Toledo no later than 1144, and 'there, seeing the abundance of books in Arabic on every subject [...] he learned the Arabic language, in order to be able to translate.'

Gerard's translations include Arabic versions of writings by Aristotle, Euclid, Archimedes, Ptolemy and Galen, as well as works by al-Kindi (*c*.801–66), al-Khwarizmi, al-Razi (*c*.854–*c*.930), Ibn Sina (Avicenna) (*c*.980–1037), Ibn al-Haytham (Alhazen) (*c*.965–*c*.1041), Thabit ibn Qurra (*c*.836–901), al-Farghani (d. after 861), al-Farabi (870–950), Qusta ibn Luqa (d. 912), Jabir ibn Hayyan (*c*.721–*c*.815), al-Zarqali (d. 1100), Jabir ibn Aflah (*c*.1120), Masha'allah (*c*.762–*c*.809), the Banu

Musa (d. second half of the ninth century) and Abu Ma'shar (d. 886). The subjects covered in these translations include 21 works on medicine; 17 on geometry, mathematics, optics, weights and dynamics; 14 on philosophy and logic; 12 on astronomy and astrology; and 7 on alchemy, divination and geomancy, or predicting the future from geographic features.

More of Arabic science was transmitted to Western Europe through Gerard than from any other source. His translations had considerable influence upon the development of European science, particularly in medicine, where students in the Latin West took advantage of the more advanced state of medical studies in medieval Islam. His translations in astronomy, physics and mathematics were also very influential, since they represented a scientific approach to the study of nature rather than the philosophical and theological attitude that had been prevalent in the Latin West. His translation of Ptolemy's *Almagest* was particularly important, for as a modern historian of science has noted, through this work 'the fullness of Greek astronomy reached western Europe.'

The Dominican monk William of Moerbeke (*c*.1220–35 – before 1286), in Belgium, was the most prolific of all medieval translators from Greek into Latin. Moerbeke claims that he undertook this translation task 'in order to provide Latin scholars with new material for study.' Moerbeke's translations included the writings of Aristotle, Archimedes, Hero of Alexandria, Ptolemy and Galen. His translations contributed to a better knowledge of the actual Greek texts of several works, and in some rare cases these constituted the only remaining evidence of Greek texts long lost, such as that of Hero's *Catoptrica*.

By the end of the twelfth century European science was on the rise, stimulated by the enormous influx of Graeco-Islamic works translated into Latin from Arabic as well as other works translated directly from Greek. By the first quarter of the following century virtually all of the scientific works of Aristotle had been translated into Latin, from Greek as well as Arabic, along with the Aristotelian commentaries of Ibn Rushd (Averroës) (1126–98). The translations included other works by both Greek and Islamic scientists on optics, geometry, astronomy, astrology, zoology, botany, medicine, pharmacology, psychology and mechanics. This body of knowledge became part of

A New Age

the curriculum at the first universities that began to emerge in the late medieval era. During the following centuries European scholars would absorb Graeco-Islamic learning and begin to make advances of their own, developing the new science that would emerge with the dawn of the new age.

CHAPTER 3
THE JAGIELLONIAN UNIVERSITY OF KRAKOW

When Nicolaus and Andreas arrived in Krakow in the autumn of 1491, the city had a population of some 20,000, a fortieth of what it is today. The city had been the capital of Poland since the reign of Casimir I, the Restorer (r. *c.*1034–58), the fourth ruler of the Piast dynasty that had been founded by Mieszko I (r. 960–92). It was devastated during the Mongol invasions that swept across Eastern Europe in 1241, 1259 and 1287 and almost destroyed the Polish kingdom, which was reunified under Wladyslaw Lokietek (r. 1306/20–33), the first king to be crowned in Krakow.

The burghers in Krakow at the time were predominately German and German-speaking. Indigenous Poles were forbidden to live in the city since princes and landlords feared the loss of manpower from their estates, though many peasants managed to settle anyway.

Krakow began to rise to prominence after the university was founded in 1364, and as capital of the Polish kingdom and member of the Hanseatic League the city began to attract craftsmen, businesses and guilds, thus becoming a centre of learning and the arts. The city also counted a large Jewish minority who were ordered out of the centre of Krakow in 1495 and resettled in the Kazimierz quarter, which was divided into Christian and Jewish sections.

The founding of the University of Krakow, along with those elsewhere in Europe, was stimulated by the assimilation of Graeco-Islamic science, first in translation from Arabic to Latin and later from the original Greek.

The earliest institution of higher learning was the university of Bologna, founded in 1088, followed in turn by Paris (*c.*1150), Oxford (1167), Salerno (1173, a refounding of the medical school), Palenzia (*c.*1178), Reggio (1188), Vicenza (1204), Cambridge (1209), Salamanca (1218) and Padua (1222), to name only the first ten. Another ten were founded in the remaining years of the thirteenth century. In the fourteenth century 25 more were founded with another 35 in the fifteenth, so that by 1500 there were 80 universities in Europe. This is an evidence of the tremendous intellectual revival that had taken place in the West, and that started with the initial acquisition of Graeco-Islamic learning in the twelfth century.

Bologna became the archetype for later universities in Southern Europe, Paris and Oxford for those in the northern part of the continent. Bologna was renowned for the study of law and medicine, Paris for logic and theology, and Oxford for philosophy and natural science. Training in medicine was based primarily on the teachings of Hippocrates and Galen, while studies in logic, philosophy and science were based on the works of Aristotle and commentaries upon them.

Although Aristotle's works formed the basis for most non-medical studies at the new universities, some of his ideas in natural philosophy, particularly as interpreted in commentaries by Averroës, were strongly opposed by Catholic theologians. One point of objection to Aristotle was his notion that the universe was eternal, which denied the act of God's creation; another was the determinism of his doctrine of cause and effect, which left no room for divine intervention or other miracles. Still another objection was that Aristotle's natural philosophy was pantheistic, identifying God with nature, which derived from the Neoplatonic interpretation of Aristotelianism by Avicenna.

This led to a decree, issued by a council of bishops at Paris in 1210, forbidding the teaching of Aristotle's natural philosophy in the university's faculty of arts. The ban was renewed in 1231 by Pope Gregory IX, who issued a bull declaring that Aristotle's works on natural philosophy were not to be read at the University of Paris 'until they shall have been examined and purged from all heresy.' The ban seems to have remained in effect for less than half a century, for a list of texts used at the University of Paris in 1255 includes all of Aristotle's available works.

Meanwhile European scholars were absorbing the Graeco-Arabic learning that they had acquired and used to develop a new philosophy of nature, which although primarily based upon Aristotelianism differed from some of Aristotle's doctrines right from the beginning.

The leading figure in the rise of the new European philosophy of nature was Robert Grosseteste (*c.*1168–1253). Born of humble parentage in Suffolk, England, he was educated at the cathedral school at Lincoln and then at the University of Oxford. He taught at Oxford and went on to take a master's degree in theology, probably at the University of Paris. He was then appointed Chancellor of the University of Oxford, where he probably also lectured on theology, while beginning his own study of Greek. When the first Franciscan monks came to Oxford in 1224 Grosseteste was appointed as their reader. He finally left the university in 1235 when he was appointed Bishop of Lincoln, his jurisdiction including Oxford and its schools.

Grosseteste's writings are divided into two periods: the first when he was chancellor of Oxford, and the second when he was bishop of Lincoln. His writings in the first period include his commentaries on Aristotle and the Bible and most of his independent treatises, while those in the second period are principally his translations from the Greek: Aristotle's *Nichomachean Ethics* and *On the Heavens*, the latter along with his version of the commentary by Simplicius, as well as several theological works.

Grosseteste's commentaries on Aristotle's *Posterior Analytics* and *Physics* were among the first and most influential interpretations of those works. These two commentaries also presented his theory of science, which he put into practice in his own writings, including six works on astronomy and one on calendar reform, as well as treatises entitled *The Generation of the Stars, Sound, The Impressions of the Elements, Comets, The Heat of the Sun, Color, The Rainbow* and *The Tides*, in which he attributed tidal action to the moon.

Grosseteste was the first medieval scholar to deal with the methodology of science, which for him involved two distinct steps. The first was a combination of deduction and induction, which he termed 'composition' and 'resolution,' a method for arriving at definitions.

The second was what Grosseteste called verification and falsification, a process necessary to distinguish the true cause from other possible causes. He based his use of verification and falsification upon two assumptions about the nature of physical reality. The first of these was the principle of the uniformity of nature, in support of which he quoted Aristotle's statement that 'the same cause, provided that it remains in the same condition, cannot produce anything but the same effect.' The second was the principle of economy, which holds that the best explanation is the simplest, that is, the one with the fewest assumptions, other circumstances being equal.

Grosseteste believed that the study of optics was the key to an understanding of the physical world, and this gave rise to his Neoplatonic 'Metaphysics of Light.' He believed that light is the fundamental corporeal substance of material things and produces their spatial dimensions, as well as being the first principle of motion and efficient causation. According to his optical theory, light travels in a straight line through the propagation of a series of waves or pulses, and because of its rectilinear motion it can be described geometrically. His researches on optics include a study of the focusing of light by a 'burning glass,' or spherical lens, which led him to predict the invention of the telescope and the microscope. 'This part of optics,' he said, 'when well understood, shows us how we may make things a very long distance off appear as if placed very close [...] and how we may make small things placed at a distance appear any size we want, so that it may be possible for us to read the smallest letters at incredible distances, or to count sand, or grains, or seeds, or any sort of minute objects.'

Grosseteste also wrote a number of treatises on astronomy. The most important of these was *De sphaera*, in which he discussed elements of both Aristotelian and Ptolemaic theoretical astronomy. He also suggests of Aristotelian and Ptolemaic astronomy in his treatise on calendar reform, *Compotus*, where he used Ptolemy's system of eccentrics and epicycles to compute the paths of the planets, though he noted that 'These modes of celestial motion are possible, according to Aristotle, only in the imagination, and are impossible in nature, because according to him all nine spheres are concentric.' Grosseteste also suggests of astrological influences in his treatise *On Prognostication*, but he later condemned astrology, calling it a fraud and a delusion of Satan.

Grosseteste's *De sphaera* was written at about the same time as a treatise of the same name by his contemporary John of Holywood, better known by his Latin name, Johannes de Sacrobosco. Sacrobosco's fame is principally based on his *De sphaera*, an astronomy text based on Ptolemy and his Arabic commentators, most notably al-Farghani. The text was first used at the University of Paris and then at all schools throughout Europe, and it continued in use until the late seventeenth century.

Grosseteste's most renowned disciple was Roger Bacon (*c*.1219–92), who acquired his interest in natural philosophy and mathematics while studying at Oxford. He received an MA either at Oxford or Paris, *c*.1240, after which he lectured at the University of Paris on various works of Aristotle. He returned to Oxford *c*.1247, when he met Grosseteste and became a member of his circle.

Bacon appropriated much of Grosseteste's 'Metaphysics of Light' as well as his mentor's emphasis on mathematics, particularly geometry. But he does go beyond Grosseteste in his commentary on Alhazen, particularly his theory of the eye as a spherical lens, basing his own anatomical descriptions on those of Hunayn ibn Ishaq (808–73) and Avicenna. Bacon also predicted the invention of wondrous machines such as self-powered ships, submarines, automobiles and airplanes, writing that 'flying machines can be constructed so that a man sits in the midst of the machine revolving some engine by which artificial wings are made to flap like a flying bird.'

Another pioneer of the new European science was Jordanus Nemorarius (fl. *c*.1220), a contemporary of Grosseteste's. Jordanus made his greatest contribution in the medieval 'science of weights' (*scientia de ponderibus*), now known as 'statics', the study of forces in equilibrium. One of the concepts he introduced was that of 'positional gravity' (*gravitas secundum situm*), which he expressed in the statement that 'weight is heavier positionally, when, at a given position, its path of descent is less oblique.' An example would be a block on an inclined plane, whose apparent weight, the force with which it presses against the surface, is greater if the angle of inclination is less. This is equivalent to resolving the weight into two components, one perpendicular to the plane, which is the apparent weight or 'positional gravity,' and the other parallel to the surface.

During the second quarter of the fourteenth century a group of scholars at Merton College, Oxford, developed the conceptual framework and technical vocabulary of the new science of motion. The most important were Thomas Bradwardine (*c.*1290–1349) and William Heytesbury (fl. 1330–9), who continued the Oxford tradition in science initiated by Robert Grosseteste.

Bradwardine's principal work is the *Tractatus Proportionum*, completed in 1328. The problem that Bradwardine tried to solve in this work was to find a suitable mathematical function for the Aristotelian law of motion, which states that the velocity (v) of an object is proportional to the power (p) of the mover divided by the resistance (r) of the medium. Bradwardine focused on the change in velocity. Heytesbury, in his *Regulae*, defined uniform acceleration as motion in which the velocity is changing at a constant rate, either increasing or decreasing. For such motion he defined acceleration as the change in velocity in a given time, which would be negative in the case of deceleration. He also introduced the notion of instantaneous velocity, for example, the speed at a particular moment, defining it as the distance travelled by a body in a given time if it continued to move with the speed that it had at that moment. He showed that, for uniformly accelerated motion, the average velocity during a time interval is equal to the instantaneous velocity at the mid-point of that interval. This was known as the Mean Speed Rule of Merton College, which was adopted by Heytesbury's successors at both Oxford and Paris.

Advances were also made at the University of Paris by Jean Buridan (*c.*1295–*c.*1358) and his student Nicole Oresme (*c.*1320–82).

Most of Buridan's extant writings are commentaries on the works of Aristotle. His most important contribution to science is his so-called impetus theory, the revival of a concept first proposed in the sixth century by John Philoponus. He explains the continued motion of a projectile as being due to the impetus it received from the force of projection, and which 'would endure forever if it were not diminished and corrupted by an opposing resistance or something tending to an opposed motion.' Buridan defines impetus as a function of the body's 'quantity of matter' and its velocity, which is equivalent to the modern concept of momentum, or mass times velocity, where mass is the inertial property of matter, its resistance to a change in its state of motion. As applied to the case of free fall, Buridan explains that gravity is not only the primary cause of the motion, but also imparts additional increments of impetus to the body as it falls, thus accelerating it, that is, increasing its velocity.

In one of Buridan's Aristotelian commentaries he asks if a proof can be given for Aristotle's geocentric model, in which the earth is at rest at the centre of the cosmos with the stars and other celestial bodies rotating around it. He notes that many in his time believed the contrary, that the earth is rotating on its axis the cosmos with the stars and other celestial bodies rotating around it. He notes that many in his time believed the contrary, that the earth is rotating on its axis and that the stellar sphere is at rest, adding that it is 'indisputably true that if the facts were as this theory supposes, everything in the heavens would appear to us just as it now appears.' In support of the earth's rotation, he says that is better to account for appearances by the simplest theory, and it is more reasonable to think that the vastly greater stellar sphere is at rest and the earth is moving, rather than the other way around. But, after refuting the usual arguments against the earth's rotation, Buridan says that he himself believes the contrary, using the argument that a projectile fired directly upward will fall back to its starting point, which is true, at least approximately, whether or not the earth is rotating.

Oresme's researches on motion are described in his *Tractus de configurationibus qualitatum et motuum*, in which he gives a graphical demonstration of the Merton Mean Speed Rule. The graph plots the velocity (v) on the vertical axis as a function of the time (t) on the horizontal axis, as, for example, in the case of a body starting from rest and accelerating so that its velocity increases by 2 feet per second every second. Graphing the motion for 4 seconds, the velocity increases each second from 0 – 2 – 4 – 6 – 8 feet per second at the end of 4 seconds. This plots in the form of a straight line rising from 0 to 8 feet per second over a time interval of 4 seconds, forming a right triangle with a height of 8 and a base of 4. The acceleration (a) is equal to the slope of the straight line, which is 8/4, or 2, in units of feet per second per second. The average velocity is half of the final velocity, which is 8/2, or 4 feet per second. The Mean Speed Rule then gives the distance travelled in 4 seconds as 4 × 4, or 16 feet. The rule can be applied over each one-second interval, so that the average velocity for each second increases from 1 – 3 – 5 – 7 feet per second. The distance in feet travelled in the first second is then 1, in the second 3, in the third 5 and in the fourth 7. These results can be generalized by the equations $v = a \times t$, and $s = a/2 \times t^2$. These are the kinematic equations formulated by Galileo in his *Dialogue Concerning the Two New Sciences* (1638), where he used Oresme's demonstration of the Merton Mean Speed Rule in his proof.

Oresme also had original ideas in astronomy, which he presented in his *Livre du ciel et du monde d'Aristote*, written in 1377 for Charles V. One of these was his comparison of the eternal motion of the celestial spheres to a perpetual mechanical clock, set in motion by God at the moment of their creation. He writes that 'it is not impossible that the heavens are moved by a power or corporeal quality in it, without violence and without work, because the resistance in the heavens does not incline them to any other movement nor to rest but only that they are not moved more quickly.'

Oresme objected to the Aristotelian notion that the earth was the stationary centre of the finite cosmos and the reference point for all motion and gravitation. He argued that motion, gravity and the directions in space must be regarded as relative, saying that God, through his omnipotence, could create an infinite space and as many universes as he chose. Oresme was thus able to reject the idea that the earth was the fixed centre of the cosmos to which all gravitational motions were directed. Instead he proposed the idea that gravity was simply the tendency of bodies to move towards the centre of spherical mass distributions. Gravitational motion was relative only to a particular universe; there was no absolute direction of gravity applying to all of space.

Oresme proposed, 'subject to correction, that the earth is moving with daily motion and the heavens not. And first I will declare that it is impossible to show the contrary by any observation; secondly from reason: and thirdly I will give reasons in favor of the opinion.'

Oresme's arguments in favour of the earth's motion would later be used by both Copernicus and Galileo. Despite all of these arguments, Oresme, who at the time had just been appointed Bishop of Lisieux, in the end rejected the idea of the diurnal rotation of the earth as being contrary to Christian doctrine, 'For God fixed the earth, so that it does not move, notwithstanding the reasons to the contrary.' Oresme's attitude was not uncommon among his clerical contemporaries, for in his position as bishop he was sworn to uphold the doctrines of the Catholic Church, even when they conflicted with his own philosophical ideas.

Meanwhile advances were being made in other areas of science, namely natural philosophy, cosmology, magnetism, astronomy and optics, as well as in mathematics and its application to astronomy and other fields of science.

The earliest extant treatise on magnetism is by Petrus Peregrinus, of whom virtually nothing is known except what appears in Peter's treatise, actually a letter, and the *Epistola Petri Peregrini de Maricourt ad Sygerum de Foucaucourt, Miltem, De Magnete* (Letter on the Magnet of Petrus Peregrinus of Maricourt to Sygerus of Foucaucourt, Soldier). Peter concluded the letter with the note that it had been 'Completed in camp, at the siege of Lucera, in the year of our Lord 1269, eighth day of August.' This would indicate that Peter was at the time in the army of Charles of Anjou, King of Sicily, who was then besieging the city of Lucera in southern Italy.

The *Epistola* is in two parts, of which the first, in ten chapters, describes the properties of the lodestone, or magnetic rock, while the second is devoted to the construction of three instruments using magnets. Peter's observations led him to make the distinction between the north and south magnetic poles; to establish the rules for the attraction and repulsion of magnetic poles; to show the magnetization of iron by bringing it in contact with a magnet; and to demonstrate that a magnetic needle when broken in half forms two separate magnets. He showed that a magnetic needle oriented itself in the north-south direction, thereby inventing the compass, which he showed could be used to map the meridians of the earth's magnetic field. He believed, mistakenly, that the poles of a magnetic needle point to the poles of the celestial sphere, the points about which the stars appear to be rotating, which are actually projections of the earth's axis of rotation. He attempted to construct a perpetual motion machine using magnets, and he blamed his failure on his lack of skill rather than the impossibility of creating an eternal source of energy. The *Epistola* was very popular in the late medieval era, as evidenced by the fact that there are at least 31 extant manuscript copies.

One of the most notable of the early European astronomers was William of St Cloud, who flourished in France during the late thirteenth century. The earliest date of his activity is 28 December 1285, when he observed a conjunction of Saturn and Jupiter, to which he refers in his *Almanach*, completed in 1292. His other major work is his *Calendrier de la reine*, also completed in 1292, dedicated to Queen Marie of Brabant, widow of Philip III, the Bold, and which he translated into French at the request of Jeanne of Navarre, wife of Philip IV, the Fair.

Queen Marie's *Calendrier* represents William's effort to put the calendar on a purely astronomical basis. This led him to contradict the computations in the ecclesiastical calendar, which he found full of errors, indicating the need for calendar reform. The purpose of his *Almanach* was to provide listings in which the positions of the celestial bodies were given directly, as contrasted to earlier tables, which only gave the elements by which those positions could be calculated. He points out the errors in the earlier planetary tables and shows how he corrected them. These were the *Toledan Tables*, used in the Muslim calendar, and those of Toulouse, the adaptation of the *Toledan Tables* to the Christian calendar. He makes no mention of the *Alfonsine Tables*, done under the patronage of King Alfonso X (1221–84) of Castile and León, which were not used in Paris before 1320.

William's observations were remarkably precise, evidence of the high level that had been achieved in European astronomy by this time. By comparing his astronomical observations with ancient Greek values he was able to measure the change in the spring equinox, which he interpreted as a steady precession rather than the trepidation theory of Theon of Alexandria, which had been revived and translated into Arabic by Thabit ibn Qurra of Harran at the *Bayt al-Hikma*.

Optics, the study of light, was another area which developed in the new European science. It began at Oxford with the work of Robert Grosseteste and his disciple Roger Bacon. The first significant advance beyond what they had achieved was by the Polish scholar Witelo (*c.*1230/35–after *c.*1275).

Witelo's most illustrious work is the *Perspectiva*, which is largely influenced by Robert Grosseteste and Roger Bacon's works as well as those of Alhazen, Ptolemy, Hero of Alexandria.

Witelo embraced the notion of 'metaphysics of light' elaborated by Grosseteste and Bacon, and said in the preface to the *Perspectiva* that visible light is an example of the propagation of the power that forms the basis of all natural causes. But he does not support Grosseteste and Bacon's theory that light rays travel from the observer's eye to the visible object. Instead Witelo follows Alhazen in holding that light rays emanate from the object to interact with the eye.

The *Perspectiva* describes experiments that Witelo performed in refraction. His method is similar to Ptolemy's in that he measured the angle of light refraction by passing air into glass and into water, for angles of incidence ranging from 10 to 80 degrees. He sought to relate the amount of refraction to the difference in the densities of the two media. He also produced the colours of the spectrum by passing light through a hexagonal crystal, observing that the blue rays were refracted more than the red.

Witelo also studied refraction in lenses by using the concept later known as the principle of minimum path, writing that 'It would be futile for anything to take place by longer lines, when it could better and more certainly take place by shorter lines.'

Witelo followed Grossteste in assuming that the 'multiplication of species' could be used to explain the propagation of any effect, including the divine emanation and astrological influences. He notes that 'there is something wonderful in the way in which the influence of divine power flows in to things of the lower world passing through the powers of the higher world.'

The next advances in optics were made by Dietrich of Freiburg (*c.*1250–*c.*1311). Dietrich, who is thought to be from Freiburg in Saxony, entered the Dominican order and probably taught in Germany before studying at the University of Paris, *c.*1275–7.

On the Rainbow and Radiant Impressions was Dietrich's major work. He was the first scholar to understand that the rainbow is caused by individual rain drops rather than the cloud as a whole. He used a glass bowl filled with water as a prototype, noting that 'a globe of water can be thought of, not as a diminutive spherical cloud, but as a magnified raindrop.' He concluded that light is refracted when entering and leaving a raindrop, and that it is internally reflected once in the primary arc of the rainbow and twice in the secondary, the second reflection reversing the order of the colours in the spectrum. Although his analysis was erroneous in a number of instances, his theory was far superior to those of any of his predecessors, and it paved the way for subsequent researches.

Dietrich's rainbow theory is quite similar to that of the Persian Kamal al-Din al-Farisi (1267–1319), which shows that the emerging European science had reached a level comparable to that of Islam science, at least in optics. But whereas the work of al-Farisi was the last great achievement of Arabic optics, Dietrich's researches would lead to the further development of European studies in the science of light, culminating in the first correct theories of the rainbow and other optical phenomena in the seventeenth century.

The Black Death of 1348 killed almost a third of Europe's population, causing a hiatus in the development of its new science. The first notable work to emerge after this catastrophe was a thesis entitled *De Docta Ignorantia* (On Learned Ignorance), published in 1440 by Nicholas of Cusa (1401–64), who in 1448 was made a cardinal by Pope Nicholas V.

Cusa believed that the universe was infinite in extent, making the idea of a centre or of a periphery meaningless. Thus the earth cannot be the centre of the universe, and since motion is relative and natural to all bodies the earth cannot be at rest. A marginal note made by Cusa in one of his manuscripts suggests that the earth cannot be fixed, but rotates on its axis once in a day and a night.

By the second half of the fifteenth century the development of European science resumed and soon rose to an even higher level. This is particularly evident in the work of Georg Peurbach (1423–61) and his student and collaborator Johannes Regiomontanus (1436–76), who in their relatively brief lives brought mathematical astronomy to a level exceeding that of Hipparchus and Ptolemy.

Therefore, by the time that Copernicus began his higher studies, the new European science was beginning to flourish, in particular at the University of Krakow. The university dated back to 1364, when King Casimir III, with permission from Pope Urban V, established a *Studium Generale* in Krakow. Krakow Academy, as it was called, was restored and modernized in 1400 by King Wladyslaw II Jagiello (r. 1384–1434), who was executing the last will and testimony of his late wife Queen Jadwiga (r. 1384–99), and thenceforth it was known as the Jagiellonian University of Krakow. From the mid-fifteenth century onward the university became one of Europe's leading academic centres of mathematics, astronomy, astrology, geography and legal studies. The geographer Hartmann Schedel, in his *Historia Mundi*, published in 1493, writes in praise of the renowned Jagiellonian University, which had a chair in astronomy and astrology as early as 1410, the first in Central Europe: 'There is in Krakow a famous university, which boasts many most eminent highly educated scholars, and in which numerous liberal arts are practiced. But the science of astronomy stands highest here, and in all Germany there is no more renowned university in this respect, as I know exactly from the narrations of many persons.'

The core of the modern university is still the fifteenth-century Collegium Maius, the Gothic building in which Copernicus studied. He and his brother Andreas entered the university in the autumn of 1491, when they were admitted at the faculty of arts. Their names are listed in the matriculation records of that year as 'Nicolaus Nicolai de Thurnia' and 'Andreas Nicolai de Thorun'. They were among 69 students who had enrolled that semester, when the total enrolment at the university was nearly 8,000.

The brothers roomed at the home of a friend of their uncle Lucas, Piotr Wapowski, whose nephew Bernard Wapowski became a lifelong friend of Copernicus. Bernard matriculated at the University of Krakow in 1493 and went on to become a renowned cartographer, noted for his maps of Poland, in which Copernicus may have collaborated with him.

The brothers remained at the University of Krakow for less than four years, leaving without taking a degree. During that time Nicolaus is known to have taken courses in mathematics, astronomy, astrology, geography and the liberal arts, his reading in the latter including Cicero, Virgil, Ovid and Seneca.

Nicolaus studied geography with Matthew of Miechow (1457–1523), author of a text that was used at universities throughout Europe. Matthew would later be the first to learn about the new heliocentric theory of Copernicus, which he came upon in 1514 in an untitled and anonymous tract, not realizing that the author was his former student.

The noted Polish astronomer Albert Brudzewski (*c*.1445–*c*.1497) was lecturing in the University of Krakow at the time and Nicolaus would have taken his course and read his works. Brudzewski may have been a pupil of Regiomontanus at the University of Vienna. He had published a commentary on Peurbach's *Theoricae novae planetarum*, in which Brudzewski formulated his own theory that the celestial orbs are not spheres but circles. He lectured on Peurbach's theory as well as Regiomontanus' *Ephemerides*.

The books on mathematics and astronomy that Copernicus is known to have studied include Euclid's *Elements*; Ptolemy's treatise on astrology, the *Tetrabiblos*; the *Alfonsine Tables*; Sacrobosco' *Sphaera*; Peurbach's *Theoricae novae planetarum* and *Tables of Eclipses;* and Regiomontanus' *Ephemerides* and *Tabulae directionum*. He had the two latter books bound together with blank pages at the back in which he copied passages from Peurbach's *Table of Eclipses*.

Brudzewski was the first to state that the moon moves in an elliptical path and always shows the same side to the earth, and he was sceptical of the geocentric planetary model of Aristotle and Ptolemy. He used the Ptolemaic model in describing the motion of the moon, adding a second epicycle.

Copernicus would later use the notion of the secondary epicycle for his lunar theory, which first appears in the *Commentariolus*, which he wrote anonymously around 1514. It is possible that he may have derived the idea from Brudzewski during his studies at the University of Krakow.

Brudzewski also made use of a mathematical method similar to one that appears in the works of the Arabic astronomers Nasir al-Din al-Tusi (1201–74) and Ibn al-Shatir (*c*.1305–*c*.1375), and to a theoretical model that Copernicus would later employ in his heliocentric theory. Recent research has shown that the works of al-Tusi and al-Shatir were also available to Copernicus when he was studying at Krakow, although they were not translated into Latin.

The works of several other Arabic scientists were available in Krakow at that time in Latin translations, including those of Masha'allah, al-Farghani, al-Kindi, al-Razi, al-Battani (858–929), Thabit ibn Qurra, Ibn al-Haytham (Alhazen) (*c*.965–1041) and Jabir ibn Aflah. Nicolaus also bought a number of books in Johann Haller's bookshop in Krakow, including a copy of the second edition of the *Alfonsine Tables* printed in Venice in 1492. These tables included some contemporary observations as well as the work of Arabic astronomers, while its mathematical basis was the Ptolemaic system of eccentrics and epicycles. He also bought copies of the astronomical tables compiled by Peurbach and Regiomontanus, indicating that he was well aware of the new science that was beginning to emerge in Western Europe, along with the heritage of Greek science that had been preserved and augmented in the Islamic world.

Copernicus was still at the University of Krakow in March 1493, when Christopher Columbus returned to Lisbon after his first voyage to the New World. Word of his voyage spread quickly through Europe and Copernicus would have learned of it that spring, but at first it was not clear what Columbus had discovered. Clarification came from the Florentine navigator and cartographer Amerigo Vespucci, who at the invitation of King Manuel I of Portugal participated as observer in several voyages of exploration along the east coast of South America in the years 1499–1502. Two letters attributed to Vespucci and describing his voyages were published early in 1503 under the title *Mundus Novus*, the New World that was soon called America after the Latinized form of his first name. Copernicus seems to have read this work soon after its publication, and he writes of the New World in Book I of *De revolutionibus*, in the section called 'How land and water make up a single globe,' where he refers to Ptolemy's treatise on geography:

> Ptolemy in his *Cosmography* extends inhabitable lands as far as the meridian circle, and he leaves that part of the Earth as unknown, where the moderns have added Cathay and other vast regions as far as 60° longitude, so that inhabited land extends in longitude farther than the rest of the ocean does [...] For reasons of geometry compel us to believe that America is situated diametrically opposite to the India of the Ganges.

Albertus Caprinus received the bachelor of arts degree from the University of Krakow in 1541. On 27 September 1542 he dedicated his *Iudicium astrologicum* (Astrological Forecast) for the year 1543. There he pays tribute to the intellectual debt that Copernicus owed to the

Jagiellonian University: 'He once enjoyed the hospitality of this city. His wonderful writings in the in the field of mathematics, as well as the additional materials he has undertaken to publish he first acquired at this university of ours as his source.'

CHAPTER 4
RENAISSANCE ITALY

Nicolaus and Andreas left Krakow early in 1496 to live with their uncle Lucas Watzenrode in the bishop's palace at Heilsberg. Lucas nominated Nicholas and Andreas to be canons of Frauenburg Cathedral, but at first his efforts were unsuccessful. Nicolaus was finally made a canon in 1497 and Andreas was elected in 1499. Both of them were elected *in absentia*, for in the autumn of 1496 Nicolaus had gone off to study in the faculty of law at the University of Bologna, where Andreas joined him two years later. Nicolaus would remain in Italy for almost seven years, except for a period of three months in 1501, when he and Andreas returned to Frauenburg to obtain an extension of their leave of absence.

At the end of the fifteenth century Italy was basically divided into three political divisions. The Kingdom of Naples comprised the southern half of the peninsula, along with Sicily and Sardinia; to its north were the Papal States, ruled by the Pope in Rome; while the northern plain was divided among the various city-states: the Republics of Siena, Florence, Genoa and Venice, and the Duchies of Ferrara and Milan.

The brothers arrived in Bologna during a brief period of peace in a series of wars between the French and the Italians that would continue at intervals for 65 years. The first of these wars had begun in September 1494, when Charles VIII of France invaded Italy with a large army. Charles had made an extravagant claim to the Neapolitan throne, part of a grandiose plan to conquer Italy and then lead a crusade to recapture Constantinople and Jerusalem. The figurehead of the crusade was to be the Turkish prince, Jem, pretender to the Ottoman throne, who was being held hostage in the Vatican by Pope Alexander VI Borgia. Charles planned to capture Jem and set up the prince as his puppet when he conquered Constantinople.

When the invasion began King Alfonso II of Naples and Pope Alexander VI were joined in an anti-French league by Florence, Siena, Bologna, Pesaro, Urbino and Imola. But the French army swept them aside and on 27 December 1494 they entered Rome, where Pope Alexander surrendered Prince Jem to Charles. Charles then went on to capture Naples on 22 February 1495, King Alfonso having fled on his approach. But a few days before Prince Jem had died of a fever, and so at the very moment of his triumph Charles lost the figurehead of his projected crusade, which he now abandoned.

Charles then withdrew his army from Italy, ending the war with a peace treaty signed at Vercelli on 10 October 1496. Charles finally reached Lyon early in November after which, according to the French statesman and historian Phillipe de Commines, 'he cared only to amuse himself and make good cheer and tourney.'

Charles died on 7 April 1498, aged only 28. Since he had no living children the throne passed to the Duke of Orleans, who succeeded as Louis XII. Louis soon resumed the war, using his claim of right as the Orleanist heir of the Visconti to invade Italy in 1499 and take Milan, along with Genoa, also using his Florentine allies to seize Pisa. This period of occupation lasted until 1504, when the remainder of the French army was forced into a humiliating withdrawal from Italy.

Despite these wars, Italy was at the height of the Renaissance during the years that Nicolaus and Andreas were studying there. The Italian Renaissance was not limited to the arts, for it also gave rise to creative geniuses in science, as in the case of Leonardo da Vinci (1452–1519). Some of these figures came from abroad, as did Copernicus and the Greek scholar Johannes Bessarion (1403–70).

Bessarion was from a family of manual labourers at Trebizond. The metropolitan of Trebizond was impressed by his intelligence and sent him to Constantinople, where he graduated from the university. Bessarion became a monk when he was 20 and joined a monastery near Mistra in the Peloponnesus, where he studied under the renowned Neoplatonist scholar George Gemistos Plethon ($c.$1355–1452). He subsequently returned to Constantinople and devoted himself to 'all the liberal arts' as well as mathematical studies. He was chosen as a delegate to the Council of Ferrara-Florence, having been appointed Metropolitan of Nicaea to give him status at the conclave. When the agreement of Union of the Greek Orthodox and Roman Catholic churches was proclaimed at the cathedral in Florence on 6 July 1439, it was read in Latin by Cardinal Cesarini and then in Greek by Bessarion.

Bessarion's experience in Florence convinced him that Byzantium could only survive in alliance with the West and by sharing in the cultural life of Renaissance Italy. Deeply discouraged by popular opposition to the Union in Constantinople, he returned to Florence late in 1440, having already become a cardinal in the Roman Catholic Church. He resided mainly at his palace in Rome from 1443 onwards, though he spent some time travelling on papal diplomatic missions and served as governor of Bologna from 1450 to 1455, but he resided in Rome. He was almost elected pope in 1455, but he lost out when his enemies warned of the dangers of choosing a Greek, and so the college of cardinals chose the Catalan Alfonso Borgia, who took the name Callistus III.

Much of Bessarion's energy was spent trying to raise military support in Europe to defend Byzantium against the Turks, but his efforts came to nought, as the Ottomans captured Constantinople in 1453 and then took his native Trebizond in 1461, ending the long history of the Byzantine Empire. Thenceforth Bessarion sought to find support for a crusade against the Turks, but to no avail.

Bessarion devoted much of his time to perpetuating the heritage of Byzantine culture by adding to his collection of ancient Greek manuscripts, which he bequeathed to Venice, where they are still preserved in the Marciana Library. Among these books, he says, were 'a Euclid and a Ptolemy, written in his own hand with the appropriate [geometrical] figures.'

The group of scholars who gathered around Bessarion in Rome included George Trapezuntios (1395–1486), whom he commissioned to translate Ptolemy's *Almagest* from Greek into Latin. Then in 1459 Trapezuntios published an attack on Platonism, suggesting that it led to heresy and immorality. Bessarion was outraged and wrote a defence of Platonism, published in both Greek and Latin. His aim was not only to defend Platonism against the charges made by Trapezuntios, but to show that Plato's teachings were closer to Christian doctrine than those of Aristotle. His book was favourably received, for it was the first general introduction to Plato's thought, which at the time was unknown to most Latins, for earlier scholarly works on Platonism had not reached a wide audience.

One of Bessarion's diplomatic missions took him to Vienna in 1460–1. At the time Vienna University was a centre of astronomical and mathematical studies following the successes of the works by John of Gmunden (d. 1442), Georg Peurbach and Johannes Regiomontanus.

Peurbach had received a bachelor's degree at Vienna in 1448 and a master's in 1453, while in the interim he had travelled in France, Germany, Hungary and Italy. He had served as court astrologer to Ladislaus V, king of Hungary, and then to the king's uncle, the emperor Frederick III. His best known works are *Theoricae novae planetarum* (New Theories of the Planets) and his *Tables of Eclipses*.

Regiomontanus, originally known as Johann Muller, took his name from the Latin for his native Königsberg in Franconia. He studied first at the University of Leipzig from 1447–50, and then at the University of Vienna, where he received his bachelor's degree in 1452, when he was only 15, and his master's in 1457. He became Peurbach's associate in a research programme that included a systematic study of the planets as well as observations of eclipses and comets.

Bessarion was disappointed with George Trapezuntios's translation of Ptolemy's *Almagest*. He asked Peurbach and Regiomontanus to produce a revised and abridged version, which they agreed to do so, as Peurbach had already begun work on a compendium of the *Almagest*, but it was left unfinished when he died in April 1461. In Peurbach's absence, Regiomontanus completed the compendium about a year later in Italy, where he had followed Bessarion. He spent part of the next four years in the cardinal's entourage and the rest in his own travels, learning Greek and searching for manuscripts of Ptolemy and other ancient astronomers and mathematicians.

Regiomontanus left Italy in 1467 for Hungary, where for four years he was a guest in the court of King Matthias Corvinus, continuing his researches in astronomy and mathematics. He then spent four years in Nuremberg, setting up his own observatory and printing press.

One of the works published before Peurbach's premature death in 1476 was *Theoricae novae planetarum*, reprinted in nearly sixty editions over three centuries. He also published his own *Ephemerides*, the first planetary tables ever printed, that gives the positions of the heavenly bodies for each day from 1475 to 1506. Columbus is said to have taken the *Ephemerides* with him on his fourth and last voyage to the New World, and to have used its prediction of the lunar eclipse of 29 February 1504 to pacify the hostile natives of Jamaica.

Regiomontanus' major work in mathematic is the *De triangulis omnimodis*, which gives a systematic method for analysing triangles. According to the historian of mathematics Carl B. Boyer, this, together with his *Tabulae directionum* (Tables of Directions), brought about 'the rebirth of trigonometry'.

The astronomical writings of Regiomontanus comprise the completion of Peurbach's *Epitome of Ptolemy's Almagest*, in which mathematical methods that were omitted in other works of elementary astronomy are now used. Copernicus must have read the *Epitome* when he was a student in Bologna, since at least two propositions influenced him in the formulation of his own planetary theory. These propositions seem to have originated with the fifteenth-century Arabic astronomer Ali Qushji, and may have been transmitted to Regiomontanus by Bessarion. If this is true, this would place Bessarion and Regiomontanus in the long chain that leads from Aristarchus of Samos to Copernicus through the Arabic and Latin scholars of the Middle Ages to the dawn of the Renaissance.

The Copernicus brothers arrived in Bologna early in the autumn of 1496, after a journey that would have taken them from six to eight weeks. At the time Bologna was ruled by Giovanni II Bentivoglio (r. 1462–1506) and had a population of 72,000. The University of Bologna was the oldest in Europe, having been founded in 1088, and received a charter in 1159 from the emperor Frederick I Barbarossa. It was renowned for its teaching of canon and civil law, in which it was the model for other European universities.

The brothers registered for the winter semester at the University of Bologna, which began on 19 October 1496. They were both in the faculty of law and enrolled in the *Natio Germanorum*, the largest of the 'nations' into which foreign students were organized at Bologna, the others being the Gallic, Portuguese, Provencal, Burgundian, Savoyard, Aragonese, Navarrese, Hungarian, Polish, Bohemian, Flemish and Italian. The reason that they were enrolled in the German 'Nation' rather than the Polish is undoubtedly because the language they spoke at home was German. The 'Records of the German Nation of Bologna University' have been preserved and published, and the statutes for 1497 note that 'We decree and ordain that in this beneficent city [Bologna] the students of canon or civil law who are of the nation of the Germans, that is, all those whose native language is German even though they may live elsewhere [...] shall be deemed and understood to be the Association of the German Nation.' An entry in the *Annales Clarissimae Nacionis Germanorum* for 6 January 1497 notes *'Dominus Nicolaus Kopperlingk de Thorn grossem novem*,' indicating that he had paid the sum of nine *grossetti* as his voluntary contribution to the association of German students. As Edward Rosen remarks: 'The contemporary value of a *grosetto* may be estimated from the fact that one such small silver coin was the price of the Nation's palm branches for Palm Sunday in the jubilee year 1500.'

The small size of Copernicus' contribution stems from the fact that he had not yet been elected to the lucrative canonry at Frauenberg Cathedral. When Copernicus was duly elected to the canonry he refused to take higher orders and thus was never a priest, despite statements to the contrary by later sources from Galileo onwards.

Although Nicolaus had matriculated at the University of Bologna as a student of law, he was free to take courses in other areas as well, for as the university regulations stipulated that 'Once matriculation had occurred, the student was free to attend courses, and the professors could not exclude him.' His first love was astronomy, for which he had acquired the requisite foundation at the University of Krakow. And so he began taking courses with Dominico Maria de Novara (1454–1504) of Ferrara, professor of astronomy at the university.

Nicolaus stayed for a time as a paying guest in Novara's house. According to Georg Joachim Rheticus (1514–74), a disciple of Copernicus, whom he calls 'my teacher,' Nicolaus collaborated with Novara in his astronomical observations. As Rheticus writes in the *Narratio prima* (First Account), an account of the Copernican theory first published in 1540: 'My teacher made observations with the utmost care at Bologna, where he was not so much the pupil as the assistant and witness of observations of the learned Domenicus Maria.'

Copernicus later described an observation he made with Novara in *De revolutionibus*, IV, 27, the section entitled 'Confirmation of What has been Expounded Concerning the Parallaxes of the Moon':

> We made this observation at Bologna, after sunset on the seventh day after the Ides of March, in the year of Christ 1497. For we observed how long the moon would occult the bright star of the Hyades (which the Romans call Paliticium) [Paliticium is the giant red star known today by its Arabic name of Aldebaran], and with this in mind, we saw the star brought into contact with the shadowy part of the lunar body and already lying hidden between the horns of the moon at the fifth hour of the night, though the star was nearer the southern horn by three quarters as it were of the width or diameter of the moon.

Novara had begun his university education at the University of Florence, where he studied under Luca Pacioli, friend and collaborator of Leonardo da Vinci. He obtained doctorates in both the arts and medicine at the University of Bologna, where one of his professors had been Giovanni Bianchini, who had corresponded with Regiomontanus. Novara taught at Bologna from 1483 until 1504, his duties, besides lecturing, including the publication of an annual almanac that gave weather forecasts for the following year, along with the date of Easter, phases of the moon, times of any eclipses, the conjunctions of the celestial bodies, his astronomical tables serving as data for astrologers. Novara would have made 21 of these prognostications, of which only about half are extant, including the one for 1498, in which he reports the occultation of Aldebaran by the moon that he and Copernicus had made the previous year.

Novara noted in his prognostication for 1489 that the latitude of Mediterranean cities, as measured by the height of the pole star above the horizon, had increased by one degree and ten minutes of arc since the time of Ptolemy. He concluded from this that a slow progressive alteration in the direction of the earth's axis was taking place, an erroneous theory that was rejected by virtually all of his contemporaries and successors. Copernicus rejected the theory too, although out of loyalty to Novara he did not mention him by name. As he writes in *De revolutionibus*, II, 6: 'But the elevations of the pole, or the latitudes of the places, and the equinoctial shadows agree with those which antiquity discovered and made note of. That would necessarily take place, since the equator depends upon the pole of the terrestrial globe.'

While Copernicus was at the University of Bologna he may have studied Greek under the very popular Antonio Urceo, known as Codro (the Pauper), who was professor of Greek at the university until his death on

11 February 1500. On 17 April 1499 Aldo Manuzio, owner of the famous Aldine Press in Venice, dedicated to Codro his edition of the Greek *Epistles of Various Philosophers, Orators, and Sophists* 'in order that you may show them to your pupils, who may thereby be inspired with a greater zeal to master more elegant literature.' As Rosen points out, this collection included the *Moral, Rustic and Amatory Epistles* of Theophylactus Simocatta, a seventh-century Byzantine writer. Copernicus translated the letters of Theophylactus from Greek to Latin, and published them in 1509, six years after his return from Italy.

While Copernicus was in Bologna he bought a copy of Giovanni Crestone's Greek–Latin dictionary, which was printed in nearby Modena on 20 October 1499. Rosen points out that 'Copernicus' well-worn copy of Crestone still survives, its margins still filled with his notes.' Copernicus would have used Crestone's dictionary in his study of Greek and in his translation of Theophylactus.

While they were still at Bologna Nicolaus and Andreas received a bank loan of 100 ducats in September 1499. The loan had been arranged through a Roman bank by Bernard Sculteti, who represented Warmia in Rome. Sculteri wrote to Bishop Lucas Watzenrode on 21 October, asking him to repay the loan that had been made to his nephews.

The last astronomical observations made by Copernicus in Bologna were done early in 1500, when he noted two conjunctions of the moon with Saturn, the first on 9 January at 2:00 A.M. and the second on 4 March at 1:00 A.M. He recorded these observations on the blank pages of his copy of the *Alfonsine Tables*.

Copernicus completed his studies in Bologna on 6 September 1500. He then went to Rome, where the only record of his presence is an observation of a partial lunar eclipse he made on 6 November 1500 at 2:00 A.M. Copernicus refers to this observation in *De revolutionibus*, IV, 14, in which he says that 'in order to determine the moon's movement in relation to the established beginnings of calendar years,' he had examined two eclipses, one that had been observed by Ptolemy in AD 136/137 and the other the one that he had seen in Rome.

Otherwise all that is known of Copernicus' stay in Rome is a statement made by Rheticus in the first chapter of the *Narratio prima*. There he says that Copernicus made an observation 'at Rome, where, about the year 1500, being 27 years of age more or less, he lectured on mathematics before a large audience of students and a throng of great men and experts in this branch of knowledge.'

However long Copernicus was in Rome during the Jubilee Year of 1500, he would almost certainly have seen Pope Alexander VI Borgia, who was then in the eighth year of his reign and at the peak of his career. Paolo Capello, writing in September 1500, says: 'The Pope is now 70 years of age; he grows younger every day; his cares never last the night through; he is always merry and never does anything that he does not like. The advancement of his children is his only care; nothing else troubles him.'

Alexander, nephew of Pope Calixtus III Borgia, had been made a cardinal by his uncle when he was only 25, and succeeded Innocent VIII as Pope on 11 August 1492, when he was 61. By that time he had fathered six children, all while he was a cardinal, and he is known to have sired two or three more during the 11 years of his papacy. The most notorious of his children was his son Cesare, whom he made a cardinal in 1493, at the age of 18, and who became Duke of Valentinois in 1498. Alexander's daughter Lucrezia, who has an unwarranted reputation as a wanton murderess, was married three times through the dynastic schemes of her father and her brother Cesare, who murdered her second husband, the Duke of Bisceglia, after which she was married off to Alfonso d'Este, who succeeded his father as the Duke of Ferrara.

Nicolaus and Andreas returned to Poland in May 1501. On 27 July they made an appeal to their chapter in Frauenburg, asking for a two-year extension of their leave so that they could complete their studies in Italy. The chapter accepted and in August they left Frauenburg for Italy, Andreas to complete his degree in canon law in Bologna and Nicolaus to study medicine in Padua. The archives in Frauenberg record that that the extension was granted to Nicolaus because 'As a helpful physician he would some day advise our most reverend bishop and also the members of the chapter.'

Nicolaus enrolled at the University of Padua in the autumn of 1501 to study medicine. Padua had come under Venetian rule in 1405, and the protection of the Serenissima enabled the university to maintain some degree of freedom and independence from the Roman Catholic Church. This was reflected in the university motto: *Universa universis patavina libertas* (Paduan Freedom is Universal for Everyone).

The university, which had been founded in 1222, was reorganized in 1399 into two divisions: a *Universitas Juristarum* for civil law, canon law and theology, and a *Universitas Aristarum* for astronomy, medicine, philosophy, dialectic, grammar and rhetoric. The student body was organized into two different groups of 'nations': the *cismontanes* for Italians, and the *ultramontanes*, for those who, like Copernicus, came from beyond the Alps.

The medical curriculum at Padua included a study of astrology, for physicians were trained to be aware of the planetary configurations when performing procedures such as blood-letting. Thus it was probably at Padua that Copernicus first read *Disputationes adversus astrologiam divinatricem* (Disputations against Divinatory Astrology) by Giovanni Pico della Mirandola (1463–94).

One of Pico's arguments against astrology was based on the fact that astronomers were still in disagreement about the relative order of Mercury and Venus among the celestial bodies in the geocentric model of Aristotle and Ptolemy, and so given these uncertainty astrological predictions were worthless. Rheticus remarks in the *Narratio prima* that Pico's criticism had become irrelevant now that Copernicus had formulated a theory that allowed the positions of the celestial bodies to be predicted accurately. 'If such an account of the celestial phenomena had existed a little before our time, Pico would have had no opportunity, in his eighth and ninth books, of impugning not merely astrology but also astronomy.'

Pico reports in the *Disputationes* that Averroës, in his *Paraphrase of Ptolemy's Almagest*, had seen a blackish spot on the sun when he was observing a conjunction of the sun and Mercury, for example, when the planet was between the earth and the sun. Only the Hebrew translation of Averroës' original version in Arabic has survived, and Pico could read Hebrew. Copernicus refers to this observation in *De revolutionibus*, I, 10, where he correctly interpreted it as a transit of Mercury, that is, when the planet is seen crossing the disk of the sun, where he cites the Arabic scholar al-Battani in saying that 'it would not be easy to see such a little speck in the midst of such beaming light.' Since Copernicus could not read Hebrew he must have learned of the observation from reading the *Disputationes* of Pico, whom he does not mention in any of his works.

Nicolaus remained at the University of Padua for only two years, and left without taking a degree, for which he would have had to study for another year. He must have continued his legal studies at Padua in addition to his courses in medicine, for in 1503 he received the degree of Doctor of Canon Law at the University of Ferrara. It has been suggested that he decided to transfer to Ferrara only because it was easier and less expensive to obtain his doctorate there than in Padua.

Their two-year extension of leave having expired, Nicolaus and Andreas left Italy in the late spring or early summer of 1503. Andreas would later go back to live in Italy, but Nicolaus did not, returning for good to the 'remote corner of the earth' where he was born, bringing the Italian Renaissance with him.

CHAPTER 5
THE BISHOPRIC OF WARMIA

After Nicolaus and Andreas returned from Italy in 1503 they took up their posts as canons of the Warmia chapter in Frauenburg. At around the same time, in either 1502 or 1504, they were joined as canons by Tiedemann Giese (1480–1550), a lifelong friend of Nicolaus who became bishop of Kulm (Chelmno) in 1538 and bishop of Warmia in 1458. Born in Danzig, he attended the universities of Leipzig and Basel and also studied in Italy.

Andreas went back to Italy in 1508 to seek medical treatment for a severe disease of the skin. Apparently he had contracted leprosy while he was in Italy, and because of this the cathedral chapter in Frauenburg excluded him on 5 October 1512. Andreas was then appointed to a post in Rome, where he died in November 1518.

Soon after his return Nicolaus joined his uncle Lucas at Heilsberg Castle (Lidzbark Warminski), the official residence of the Bishops of Warmia, about 40 miles southeast of Frauenburg. He remained at Heilsberg for seven years, serving as secretary of state and personal physician to his uncle Lucas Waczenrode, Bishop of Warmia. Nicolaus also participated in the local Prussian diets, or parliaments, as a member of the group of canons called the Chapter of Warmia.

On 1 January 1504 Nicolaus went to Marienburg (Malbork) to attend a meeting of the Land Diet of the Prussian States, at which Bishop Lucas presided. The assembly decided to convene another meeting 20 days later at Elbing (Elblag), where the Prussian States solemnly refused to make a pledge of loyalty to King Alexander, grandson of Wladyslaw II Jagiello, who had become Grand Duke of Lithuania in 1492 and succeeded his childless elder brother John I Albert as King of Poland in 1501, reuniting the two states.

In May 1504 Nicolaus went with Lucas to a meeting of the Prussian delegates with King Alexander at Thorn, after which they accompanied the King to Danzig. He attended another assembly of the Prussian States with his uncle in 1506 from 20 August to 15 September. Nicolaus probably also accompanied Lucas to Krakow on 24 January 1507 to attend the coronation of Sigismund I, who had succeeded his brother Alexander as King of Poland the previous year.

On 24 January 1507 Nicolaus was appointed Personal Physician of the Bishop of Warmia, with a salary of 15 marks a year in addition to his income as a canon, which in 1519 was 98 marks. Soon afterwards Bishop Lucas became seriously ill, but Nicolaus nursed him back to health, and he accompanied his uncle to an assembly of the Prussian States from 1–4 September of that year, the last one they would attend together.

Copernicus also served as physician of the Warmia chapter, treating his fellow canons along with local clerics and dignitaries as well as the townspeople of Frauenburg and other places. In 1526 he accompanied King Sigismund to Danzig as his personal physician, and he also gave medical advice to Duke Albrecht of Prussia, Grand Master of the Teutonic Knights.

Copernicus is known to have had at least eight medical books in his possession, most of which he probably acquired when he was at the University of Padua. These included two works by Michele Savaranola (*c.*1384–1468), grandfather of Friar Girolamo Savoranola, the firebrand Florentine preacher, one of them a treatise on fevers and the other a handbook of practical medicine. The others were a dictionary of medical simples by Matteo Silvatico, published in 1498; a practical encyclopaedia of medicine by Arnold of Villanova, published *c.*1485; a work on fevers with a commentary on Avicenna, by Ugo Benzi (1376–1439); a book on practical medicine by Velescusv of Taranta, a Portuguese physician, published in 1490; a book on surgery by Pietro d'Argelatta, published in 1499; and the *Materia medica* of Dioscorides, which had been given to him by his colleague Felix Reich.

One of the prescriptions used by Copernicus was a pill made of 1 scruple of 16 medicinal plants with 5 grains of rhubarb, along with

1 scruple of aloe syrup compounded with the essence of violets or roses. The pill was supposed to cure a variety of problems: grey hair, poor vision, poor digestion, catarrh, cough, stomach gas, intellectual weakness, poor nerves, plague, gout, insomnia and colic.

In 1509 Copernicus published his first work, a Latin translation published by the Aldine press of the fictitious epistles of the seventh-century Byzantine writer Theophylactus Simocatta. The 85 letters are arranged in triads. In each group of three the first letter is ethical in character, the second is rustic and the third concerns love. Letter No. 85, which comes after the 28th triad, is ethical, and recommends a stroll through a cemetery, where 'You will behold man's greatest joys as in the end they take on the lightness of dust.'

He dedicated the book to his uncle Lucas, noting that the love letters could just as well be called moral epistles. He also remarked that Theophylactus should be congratulated for giving succinct rules of conduct, combining 'the gay with the serious, and the playful with the austere, that every reader may pluck what pleases him most in these letters, like an assortment of flowers in a garden. Yet Theophylactus disposed so much of value in all of them that they seem to be not letters but rather laws and rules for the conduct of human life.'

The first letter gives a good idea of the literary quality of the work:

1. Ethical. From CRITIAS to PLOTINUS
   The cricket is a musical being. At the break of dawn it starts to sing. But much louder and more vociferous, according to its nature, it is heard at the noon hour, because intoxicated by the sun's rays. As the songster chirps, then it turns the tree into a platform and the field into a theatre, performing a concert for the wayfarers.

Despite its lack of literary merit, this work is evidence that the revival of classical learning had penetrated even into areas that were far from centres of culture. Edward Rosen writes: 'This slender volume was the first independent translation of a Greek author to be printed in Poland. It constituted Copernicus' modest contribution to the spread of the humanist movement in his native region.'

Copernicus also translated into Latin another epistle in the Aldine collection. This fictitious letter from Lysis the Pythagorean to Hipparchus was originally inserted in the *Revolutions* by Copernicus, but on later reflection he deleted it.

Copernicus has been credited with a poetic work called *Septem sidera* (Seven Stars), in the form of seven odes with seven verses each, which was discovered and published in 1629 by Jan Brozek. Jozef Rudnicki, writing in 1943, gives a fanciful interpretation of this work, which more recent scholarship no longer attributes to Copernicus.

About this time Nicolaus left his uncle's service at Heilsberg and rejoined the cathedral chapter at Frauenburg, where he would spend most of the remainder of his life. It may be that by then he had decided to return to the astronomical research he had begun in Italy, which he would not have the time to do while serving as his uncle's minister of state.

The earliest observation recorded in *De revolutionibus* after Copernicus returned from Italy describes a lunar eclipse he saw in 1509, probably in Heilsberg or possibly Krakow, and he observed another lunar eclipse in 1511 in Frauenburg. Copernicus notes that Frauenburg has the same longitude as Krakow. Frauenburg is actually about one quarter of a degree west of Krakow, but the astronomical instruments used by Copernicus were not nearly sensitive enough to measure this small difference. Copernicus referred to Frauenburg in Greek as Gynopolis, or 'Women's Town,' as he writes in saying that it lies on the same meridian as Krakow: 'All this with reference to the Krakow meridian, since Gynopolis, commonly called Frauenburg, where we took our observations at the mouth of the Vistula, lies under this meridian, as the eclipses of the sun and moon observed in both places at the same time teach us.'

There are records of about 60 observations made by Copernicus, of which 27 are cited in *De revolutionibus*. But Swerdlow and Neugebauer, quoting Copernicus, note that the observations used in *De revolutionibus* were selected from many more. They conclude that 'the number of observations taken for possible use is nowhere near 27 or 60, but more like a few hundred, and in addition to these there are observations of unusually close conjunctions or alignments of planets and stars that he could not but notice.'

At the beginning of the sixteenth century Frauenburg had a population of 1,200, about half of what it is today. Frauenburg is on the shore of Frisches Haff, known today as Vistula Bay, the long lagoon on the Vistula delta east of Danzig that stretches from Elbing on the west to Köningsberg on the east.

The Bishopric of Warmia

This region is described in *Le Miroir du Monde* published by the Dutch writer Abraham Ortelius at Amsterdam in 1598 in his account of Prussia:

> On the east is Lithuania, on the south is Poland, on the west is Pomerania, and on the north is Livonia and the Oriental [Baltic] Sea. There are several good harbours, and amber can be found on the shores of the sea. It abounds in all sorts of grains, game animals, and fishes. There are several beautiful and great cities, among which Danzig is a big merchant city on the coast, at the mouth of the Wixel [Vistula] river. Then there are Elbing, and Koenigsberg [Köningsberg] where is the court of the Prince. The country is quite well populated. Everywhere in the cities and along the coast and along the coast German is spoken. But in the country and the villages, the old accustomed language is still spoken.

Frauenburg began its history in the thirteenth century as a Prussian defensive stronghold, where the first Bishop of Warmia took his seat in 1242. According to tradition, when the last Prussian lord of the stronghold died his widow Gertrude offered the fortress to the bishop, who named it Frauenberg, 'Our Lady's Fortress.' It is more likely that the name comes from a fortified chapel or monastery of the Virgin, which in the fourteenth century was replaced by a splendid Gothic cathedral. On 8 July 1319 Bishop Eberhard of Neisse granted the town the same Lübeck city rights that had been given to many members of the Hanseatic League; in this document it is simply called *Civitas Warmiensis*, or Warmian City.

Frauenburg was sacked and burned in 1414 during a conflict between the Teutonic Knights and Poland known as the Hunger War. By the Second Peace of Thorn, signed in 1466, Frauenburg had become an important town of the Prince-Bishopric of Warmia and part of the Polish province of Royal Prussia.

Early in 1512 Nicolaus accompanied his uncle to Krakow to attend the wedding of King Sigismund to Barbara Zapolya, who was crowned as Queen of Poland. A short poem celebrating the marriage has been attributed to Copernicus. On the way home Bishop Lucas died in Thorn on 29 March 1512, after which his body was brought back to Frauenburg and buried in the cathedral. He was replaced by Fabian von Lossainen (Luzjanski), a former classmate of Nicolaus at the University of Bologna, who was elected bishop of Warmia on 5 April. But the choice was not approved by King Sigismund until 7 December of that year, when it was agreed that in future elections the king would nominate the four possible candidates. Nicolaus, who had been named chancellor of the cathedral chapter, wrote a letter to Pope Julius II defending the King's right to nominate a successor.

Frauenburg is still dominated by the Gothic cathedral, dedicated to the Ascension of the Virgin and St Andrew Apostle, built in the years 1329–88 on the eminence known as Cathedral Hill. The bishop's mansion was adjacent to the cathedral, built into the defence walls that surrounded the acropolis on Cathedral Hill. When Copernicus returned to Frauenburg he was one of 16 canons in the cathedral chapter of Warmia. The canons lived in Frauenburg, other than two or three who resided elsewhere in Warmia or in Rome as representatives of the chapter. Most of the canons lived in a dormitory on the other side of the cathedral. Each of them also had a *curia*, a house in the town outside the walls on Cathedral Hill, and some also had a villa in the countryside.

In 1514 Copernicus bought a house just outside the west gate of Cathedral Hill. The previous year he had moved out of the dormitory inside the fortress and took up residence within the defence tower in the northwest corner of the fortress walls, which he had purchased from the cathedral chapter. The tower had three floors, the uppermost of which had windows on all sides, so that he could look out over the town and the surrounding countryside. Nearby he also built an observatory in the form of a viewing platform, where he set up his astronomical instruments with an unobstructed view of the celestial sphere. The observatory was built at his own expense, as recorded in the chapter archives, which note that, in April 1513, he paid into the treasury of the chapter money for 800 bricks and a barrel of chlorinated lime from the cathedral work-yard.

In *De revolutionibus*, Copernicus describes three instruments that he may have used in his observations. These are the quadrant, the triquetrum and the armillary sphere. All three instruments date back to antiquity, and are described by Ptolemy in the *Almagest*. Using these instruments and comparing his measurements with those of Ptolemy, Copernicus was able to show that the tilt of the earth's axis varies in time periodically, concluding that 'it was never greater than 23°52' and will not ever be less than 23°28'.'

The triquetrum, also known as Ptolemy's rulers, is described by Copernicus in *De revolutionibus*, IV, 15, the section entitled 'Construction of the instrument for observing parallaxes.' This instrument, which he used to measure the parallax of the moon, is based on Ptolemy's description in the *Almagest*, IV, 12 of the instrument he used for the same purpose, which he called the parallel acticon.

Copernicus reports two observations of the moon that he made with the triquetrum, one in 1522 and the other in 1524, both in Frauenburg. The results of these observations enabled him to compute the lunar parallax, which he compared with the values found by Ptolemy and other Greek astronomers. He concluded that 'the difference [in his two observations of the lunar zenith distance] of 1°5' came from the lunar parallax, which according to Ptolemy should have been 1°38' and also according to the theory of the ancients.'

The third of the three instruments, the armillary sphere is described in *De revolutionibus*, IV, 14, where Copernicus, following Ptolemy in the *Almagest*, IV, 1, describes it as an astrolabe, though he should more properly called it a spherical astrolabe.

The first public recognition of the growing reputation of Copernicus as an astronomer came in 1514, in connection with the programme of calendar reform that had emerged during the Fifth Council of the Church, held at the Lateran Palace in Rome in the years 1512–17. The Council had been summoned by Pope Julius II (r. 1503–13) and continued by his successor Leo X (r. 1503–21).

While the Council was in session Pope Leo announced that he had 'consulted the greatest experts in theology and astronomy,' whom he had 'advised and encouraged to think about remedying and suitably correcting' the calendar. He went on to say that 'they have conscientiously heeded me and my instructions, some of them in writing, some orally.'

But when these experts failed to produce any solutions to the problem, Leo looked farther afield for help. On 21 July 1514 he sent a message to Maximilian I, the Holy Roman Emperor, requesting that 'of all the theologians and astronomers who you have in your empire and domains, you should order [...] every single one of high renown [...] to come to this sacred Lateran Council [...] But if there be any who for a legitimate reason cannot come to the Council, Your Majesty will please instruct them [...] to send me their opinions carefully written.'

A similar message, dated 24 July 1514, was sent to the heads of governments and universities throughout Europe, and this was repeated twice during the next two years.

The Pope had appointed Paul of Middleburg (1445–1533), Bishop of Fossombrone, as head of the commission for calendar reform. Paul had been professor of mathematics at the University of Padua, and he had also served the Duke of Urbino as his physician and astrologer. He had written a treatise on the proper date for the celebration of Easter, published at Fossombrone in 1513, where he refers to an unnamed booklet, which may have been by Copernicus. As Leo wrote to Paul on 16 February 1514: 'I have great need of your ability and erudition in computing and investigating chronological matters related to the Roman calendar as well as in the items on the agenda of the sacred Lateran Council. I therefore urge you to come to Rome at the earliest time convenient to you, for your presence here is important to me.'

Paul, in a report to the Pope written in 1516, lists the name of Copernicus among the experts who had written in response to Leo's request for help in reforming the calendar, but not among those who had actually come to Rome. Otherwise there is no record of the correspondence between Paul and Copernicus. Copernicus mentions Paul in the preface of *De revolutionibus*, addressing Pope Paul III:

> For not many years ago under Leo X when the Lateran Council was considering the question of reforming the Ecclesiastical Calendar, no decision was reached, for the sole reason that the magnitude of the year and the months and the movements of the sun and moon had not been measured with sufficient accuracy. From that time on I gave attention to making more exact observations of these things and was encouraged to do so by that most distinguished man, Paul, Bishop of Fossombrone, who had been present at these deliberations.

Then, in *De revolutionibus*, III, 16, Copernicus writes of 'the ten or more years in which we applied our intelligence to these things [i.e., the length of the year] and especially in the year of Our Lord 1515 [...] and 178 days 53 and half minutes [from the autumnal equinox] to the spring equinox.' He had also noted, in *De revolutionibus*, III, 14, referring to the Arabic scholar Thabit ibn Qurra, that 'We find that the magnitude of the year and its equality is only 1 second 10 thirds greater than that of Thebith ben Chora recorded it to be, so that it contains

365 days [...] 6 hours 9 minutes 40 seconds, and its fixed equality with reference to the fixed stars is disclosed.'

Thus it would appear that Copernicus has begun to work on *De revolutionibus* as early as 1515. The central idea of this revolutionary work, that the sun and not the earth is the centre of the planetary system, seems to have come to him even earlier, as evidenced by an untitled and anonymous work that was later given the title *Nicolai Copernici de hypothesibus motuum caelestium a se constitutis commentariolus* (Nicholas Copernicus, Sketch of his Hypotheses for the Celestial Motions). This came to be known as the *Commentariolus* (Little Commentary) the first notice of the new astronomical theory that Copernicus had been developing. He seems to have given copies of this short handwritten treatise to a few friends but never published it in book form. One of the copies may have come into the hands of Paul of Middleburg and prompted him to contact Copernicus about calendar reform.

Only two manuscript copies of the treatise have survived, the first of which was published in Vienna in 1878 and the second in Stockholm in 1881. The earliest record of the treatise is a note made by Matthew of Miechow, Copernicus' former teacher at the University of Krakow. Matthew noted in a catalogue of the books in his library, dated 1 May 1514, that one of them was 'a manuscript of six leaves expounding the theory of an author who asserts that the earth moves while the sun stands still.' Matthew was unable to identify the author of this treatise, since Copernicus had not written his name on the manuscript. But there is no doubt that the manuscript was by Copernicus, because the author made a marginal note saying that he reduced all his calculations 'to the meridian of Krakow, because [...] Frauenburg [...] where I made most of my observations [...] is on this meridian as I infer from lunar and solar eclipses observed at the same time in both places.' Simon Starowolski (1588–1656), one of the first biographers of Copernicus, remarked that 'as is clear from [lost] letters written with his own hand, Copernicus conferred about eclipses and observations of eclipses with Krakow mathematicians, formerly his fellow students.'

Copernicus seems to have given a copy of this work to his disciple Rheticus in 1539, four years before he died. When Rheticus died in 1574 he bequeathed the tract and all of his other books to his colleague Thaddeus Hajek, personal physician to the Holy Roman Emperor Rudolph, having previously informed him that the author of the anonymous work was Copernicus. The following year Hajek met the young Danish astronomer Tycho Brahe, who told him of the high regard he had for Copernicus. Hajek then gave the tract to Brahe, who later, in his *Astronomiae instauratae progymnasmata*, wrote of how he acquired it:

> A certain little treatise [tractatulus] by Copernicus concerning the hypotheses which he formulated was presented to me in handwritten fo[rm] some time ago at Regensburg by that distinguished man Thaddeus Hajek, who has long been my very close friend. Subsequently I sent the treatise to certain other astronomers in Germany. I mention this fact to enable the persons into whose hands the manuscript comes to know its provenance.

Edward Rosen, in a footnote to his edition of the *Commentariolus* in *Three Copernican Treatises*, concludes that 'its date of composition is narrowed down to the dozen years between July 15, 1502 and May 1, 1514.' The first of these dates is the time of publication of an astronomical work referred to by Copernicus in the *Commentariolus*, while the second is the date of the catalogue of Matthew of Miechow.

Elsewhere Rosen revises his estimate of the earliest possible date of composition of the *Commentariolus* to the latter half of 1508, since it is already a mature work that seems to be the result of some years of careful development. Thus the *Commentariolus* may have come into being just before Copernicus left the service of his uncle Lucas and returned to Frauenburg. For there he would have had more time to develop his revolutionary theory, the central idea of which he states as an assumption at the very beginning of the *Commentariolus*, that 'the sun is the center of the universe.'

CHAPTER 6
THE LITTLE COMMENTARY

The *Commentariolus*, or *Little Commentary*, was not printed during the lifetime of Copernicus. Copies of the document were circulated among scientists, including the Danish astronomer Tycho Brahe, but then it disappeared for three centuries. It was first published in 1878 after a copy of the manuscript had been found in Vienna. An annotated English translation was published in 1939 by Edward Rosen in his *Three Copernican Treatises*. The other two manuscripts were the *Letter Against Werner* and the *Narratio prima*, the latter written by Copernicus' disciple Georg Joachim Rheticus.

The *Commentariolus* is subtitled 'Nicholas Copernicus sketch of his hypotheses for the heavenly motions,' which begins with a discussion of the ancient Greek model of the concentric celestial spheres. He goes on to write about Ptolemy's theory to explain the apparently irregular motion of the planets through the use of a combination of eccentric circles, and epicycles, an approach that he thought to be ultimately unsatisfactory, particularly because of the introduction of the equant: 'For these theories were not adequate unless certain equants were also conceived; it then appeared that a planet moved with uniform velocity neither on its deferent nor about the center of its epicycle. Hence a system of this sort seemed neither sufficiently absolute nor sufficiently pleasing to the mind.'

He then writes of how these failings of the existing model eventually led him to formulate a simpler and more satisfactory planetary theory, provided that he made certain assumptions:

> Having become aware of these defects, I often considered whether there could perhaps be found a more reasonable arrangement of circles, from which every apparent inequality would be derived and in which everything would move uniformly about its proper center, as the rule of absolute motion requires. After I had addressed myself to this very difficult and almost insoluble problem, the suggestion at length came to me how it could be solved with fewer and much simpler constructions than were formerly used, if some assumption (which are called axioms) were granted me.

The axiomatic assumptions, eight in number, are all stated simply and clearly. The first is that the celestial spheres do not all have the same centre. The second is that the earth is not the centre of the universe, but only of the gravitational force and of the lunar sphere. This assumption runs counter to both Aristotle's geocentric model doctrine and his theory of natural place and motion, the latter holding that earthly bodies gravitate towards their natural place at the centre of the cosmos, the earth's centre.

The most revolutionary assumption is the third, which states that all of the celestial spheres, for instance that of the fixed stars as well as those of the planets, including the earth, revolve around the sun, 'the center of the universe.' He follows this with the assumption that the radius of the earth's orbit around the sun is negligible compared to the distance of the fixed stars. Archimedes had made this same assumption in his *Sand-Reckoner* in explaining why the stars do not exhibit parallax, that is, appear to move in the outermost celestial sphere when viewed from the earth at different places in its orbit around the sun in the heliocentric theory proposed by Aristarchus of Samos.

This suggests that the idea of a sun-centred universe had come to Copernicus from reading *The Sand-Reckoner*. The examination of this possibility, first by Edward Rosen and then by Owen Gingerich, is a fascinating detective story, leading through a paper trail of documentary evidence dating from antiquity to the dawn of the Renaissance.

Copernicus refers to Aristarchus of Samos four times in the final version of *De revolutionibus*, but three of them are mistaken attributions.

The fourth reference includes Aristarchus in a list of astronomers who thought that the year was exactly 365¼ days long. But nowhere does he mention that Aristarchus had in the mid-third century BC proposed that the sun and not the earth was the centre of the cosmos. Copernicus had referred indirectly to the heliocentric theory of Aristarchus in his original manuscript, but deleted it from the version of *De revolutionibus* printed in 1543. The suppressed passage, which had been in the last paragraph of Book I, Chapter 11, contains this phrase: 'Philolaus believed in the mobility of the Earth and some even say that Aristarchus of Samos was of that opinion.' Rosen and then Gingerich reached the same conclusion, which is that Copernicus conceived the heliocentric theory independently of Aristarchus. As Gingerich writes: 'There is no question but that Aristarchus had the priority of the heliocentric idea. Yet there is no evidence that Copernicus owed him anything. As far as we can tell, both the idea and its justification were found independently by Copernicus.'

The last three of the seven assumptions made by Copernicus in the *Commentariolus* concern the apparent motions of the stars, sun and planets, which he attributes to the motions of the earth.

> 5. Whatever motion appears in the firmament arises not from motion of the firmament, but from the earth's motion. The earth together with its circumjacent element performs a complete rotation on its fixed poles in a daily motion, while the firmament and highest heaven remains unchanged. 6. What appear to us as motions of the sun arise not from its motion but from the earth and our sphere, with which we revolve around the sun like any other planet. The earth has, then, more than one motion. 7. The apparent retrograde and direct motion of the planets arises not from their motion but from the earth's. The motion of the earth, alone, therefore, suffices to explain so many apparent inequalities in the heavens.

The last assumption should be qualified, since the apparent motion of the planets is the resultant of their own orbital motion around the sun as well as those of the earth.

After stating these assumptions, Copernicus goes on to say that 'I shall endeavor briefly to show how uniformity of the motions can be saved in a systematic way. However, I have thought it well, for the sake of brevity, to omit from this sketch mathematical demonstrations, reserving these for my larger work.'

Rosen, in a footnote to the passage above, writes that 'From this reservation we may infer that when Copernicus wrote the *Commentariolus* he had already planned *De revolutionibus* or was at work upon it.'

The section after the assumptions is entitled 'The Order of the Spheres,' counting inward from that of the fixed stars through those of the planets including the earth, which is the centre of the sphere of the moon. The period of revolution of each planet is given as well, decreasing with the radius of their orbit. The value given for each planet is its sidereal period, that is, the time it takes to complete one revolution with respect to the stars, as distinguished from the synodic period, the time it takes a planet to complete one period as observed from the earth. Here Copernicus had to compute each sidereal period from the synodic period given in the planetary tables.

> The celestial spheres are arranged in the following order. The highest is the immovable sphere of the fixed stars, which contains and gives position to all things. Beneath it is Saturn, which Jupiter follows, then Mars. Below Mars is the sphere on which we revolve; then Venus; last is Mercury. The lunar sphere revolves about the center of the earth and moves with the earth like an epicycle. In the same order also, one planet also surpasses another in speed of revolution, according as they trace greater or smaller circles. Thus Saturn completes its revolution in thirty years, Jupiter in twelve, Mars in two and one-half, and the earth in one year; Venus in nine months and Mercury in three.

The period given for Mars is too large, and in *De revolutionibus* Copernicus reduced it to two years, bringing it closer to the true value of 687 days, or one year and ten and one-half months.

The next section is on 'The Apparent Motions of the Sun.' Here Copernicus says that the sun, as observed from the earth, has three motions, the first being its motion along the ecliptic in an eccentric circle, in other words, one whose centre is displaced from the centre of the earth:

> First it revolves annually in a great circle about the sun in the order of signs, always describing equal arcs in equal times; the distance from the center of the circle to the center of the sun is 1/25 of the radius of the circle. The radius is assumed to have a length imperceptible in comparison with the height of the firmament; consequently the sun

appears to revolve with this motion, as if the earth lay in the center of the universe. However, this appearance is caused by the motion not of the sun but of the earth, so that, for example, when the earth is in the sign of Capricornus, the sun is seen diametrically opposite in Cancer, and so on.

Copernicus goes on to say that 'The second motion, which is peculiar to the earth, is the daily rotation on the poles in the order of the signs, that is from west to east. On account of this rotation the entire universe appears to revolve with great speed. Thus does the earth rotate with its circumadjacent waters and encircling atmosphere.'

The third motion concerns the tilt of the earth's axis of rotation, that is its inclination to the perpendicular axis of the ecliptic plane, which in the time of Copernicus was about 23½°. As Hipparchus had discovered, the axis precesses very slowly around the ecliptic's perpendicular, tracing out the surface of a cone. Copernicus was aware of this, but he complicated his theory by introducing another conical motion to keep the earth's axis pointing in the same direction while the crystalline sphere in which it was embedded rotated annually. The period of this supposed motion he took to be slightly different than the time it takes the earth to rotate around the sun, the difference being due to the very slow motion of the equinoxes in the opposite direction. The latter motion is known as the precession of the equinoxes, since in effect it causes the equinoxes (and solstices) to occur slightly earlier each year.

The next section is entitled 'Equal Motion Should Be Measured Not by the Equinoxes but by the Fixed Stars.' Copernicus points out that measurements of the length of the year made by Ptolemy and other earlier astronomers have been in reference to the solstices and equinoxes. Because of the precession of the equinoxes, he insists that measurements of celestial time intervals should be sidereal, that is, referring to the fixed stars, as he had done in determining the length of the year.

The following section deals with the moon, which has two inequalities in its motion. Thus Copernicus is forced to resort to a secondary epicycle, a method he may have derived from Albert Brudzewski at the University of Krakow.

The rest of the *Commentariolus* is concerned with planetary motion, first the superior planets, those in orbit beyond the earth – Saturn,

Jupiter and Mars – then the inferior – Venus and Mercury – whose orbits are inside that of the earth.

> Saturn, Jupiter and Mars have a similar system of motions, since their deferents completely enclose the great circle and revolve in the order of signs about its center as their common center. Saturn's deferent revolves in 30 years, Jupiter's in 12 years, and that of Mars in 29 months [one month less than the value he gave earlier]; it is as though the size of the circles delayed the revolutions. For if the radius of the great circle is divided into 25 units, the radius of Mars' deferent will be 38 units, Jupiter's 130, and Saturn's 230. By 'radius of the deferent' I mean the distance from the center of the deferent to the center of the first epicycle.

He goes on to describe the system of epicycles that he used for the superior planets, where, as he explains later, 'the radius of the first epicycle in each case is three times as great as the second.'

The places where the planet is closest and farthest to the centre of the first epicycle are called the planetary apsides. Copernicus, following Ptolemy, believed that the celestial coordinates of these apsides were constant. Later he discovered that the apogees changed in time so that the apsides were not invariable, and in *De revolutionibus* he corrected his theory.

Copernicus then writes of how the apparent retrograde planetary motion, which he calls the 'second inequality,' can be explained in the case of the superior planets.

> There is a second inequality, on account of which the planet seems from time to time to retrograde, and often to become stationary. This happened by reason of the motion, not of the planet, but of the earth changing its position in the great circle. For since the earth moves more rapidly than the planet, the line of sight directed towards the firmament regresses, and the earth more than neutralizes the motion of the planet. This regression is most notable when the earth is nearest to the planet, that is, when it comes between the sun and the planet at the evening rising of the planet. On the other hand, when the planet is setting in the evening or rising in the morning, the earth makes the observed motion greater than the actual. But when the line of sight is moving in the direction opposite to that of the planets and at an equal rate, the planets appear to be stationary, since the opposed motions neutralize

each other; this commonly occurs when the angle between the sun and the plane is 120°.

Copernicus then explains how the magnitude of the second inequality varies for the three superior planets:

> In all cases the lower the deferent on which the planet moves, the greater is the inequality. Hence it is smaller in Saturn than in Jupiter, and again greater in Mars, in accordance with the ratio of the great circle to the radii of the deferents. The inequality attains its maximum for each planet when the line of sight for the planet is tangent to the circumference of the great circle. In this manner do these three planets move.

Copernicus goes on to discuss the variation in the observed longitude of the superior planets, that is, their angular position north or south of the ecliptic:

> In latitude they have a twofold deviation. While the circumferences of the inclinations lie in a single plane with their deferent, they are inclined to the ecliptic. This inclination is governed by the inclination of their axes, which do not revolve as in the case of the moon, but are directed always towards the same region of the heavens. Therefore the intersections of the deferent and the ecliptic (these points are called the nodes) occupy eternal places in the firmament.

Copernicus had derived from Ptolemy the idea that the nodes, like the apsides, are invariable. But having discovered the motion of the apsides, in *De revolutionibus* he corrected his view that the nodes were fixed.

He completes his discussion of the superior planets by showing how their celestial latitudes change depending on the position of the earth on the ecliptic, as can be demonstrated using a deferent and two epicycles:

> The motion of the earth in the great circle also causes the observed latitudes to change, its nearness or distance increasing or diminishing the angle of the observed latitude, as mathematical analysis demands. This motion in libration occurs along a straight line, but a motion of this sort can be derived from two circles. These are concentric, and one of them, as it revolves, carries with it the inclined poles of the other. The lower

circle revolves in the direction opposite to that of the upper, and with twice the velocity. As it revolves it carries with it the poles of the circle which serves as deferent to the epicycles. The poles of the deferent are inclined to the poles of the circle halfway above at an angle equal to the inclination of these poles to the poles of the highest circle. So much for Saturn, Jupiter and Mars and the spheres which enclose the earth.

He now goes on to describe the motion of the inferior planets, beginning with Venus. Here he uses a system with a deferent and two epicycles, whereas later, in *De revolutionibus*, he changes to one employing a combination of two eccentric circles.

Copernicus then explains how the apparent retrograde motion of Venus can be determined by the orbital motion of the earth, which for the case of the inferior planets is slower than their orbital velocity. He also points out why Venus is never in opposition to the sun, that is, diametrically opposite to the sun in the celestial sphere, which is of course also the case for Mercury.

> Venus seems at times to retrograde, particularly when it is nearest to the earth, like the superior planets, but for the opposite reason. For the regression of the superior planets happens because motion of the earth is more rapid than theirs, but with Venus it is slower; and because the superior planets enclose the great circle, whereas Venus is enclosed within it. Hence Venus is never in opposition to the sun, since the earth cannot come between them, but moves within fixed distances on either side of the sun. These distances are determined by tangents to the circumference drawn from the center of the earth, and never exceed 48° in our observations. Here ends the treatment of Venus' motion in longitude.

He then treats the variation of Venus in observed latitude, its angular distance north and south of the celestial equator:

> Its latitude also changes for a two-fold reason. For the axis of the deferent is inclined at an angle of 2½°, the node whence the planet turns north being in the apse. However, the deviation which arises from this inclination, although it is one and the same, appears twofold to us. For when the earth is on the line drawn through the nodes of Venus, the deviations on one side are seen above, and on the opposite side below; these are called the reflexions. When the earth is a quadrant's distance

from the nodes, the same natural inclinations of the deferent appear, but they are called the declinations. In all the other positions of the earth, both latitudes mingle and are combined, each in turn exceeding the others; by their likeness and difference they are mutually increased and eliminated.

The last section is devoted to Mercury, which Copernicus says is the most difficult of all the celestial bodies to study and observe: 'Of all the orbits in the heavens the most remarkable is that of Mercury, which traverses almost untraceable paths, so that it cannot easily be studied. A further difficulty is the fact that the planet, following a course generally invisible in the rays of the sun, can be observed for a very few days only. Yet Mercury too will be understood, if the problem is attacked with more than ordinary ability.'

As in the case of Venus, Copernicus originally tried to describe the motion of Mercury with a deferent and two epicycles. But this combination of epicycles proved to be inadequate, for it did not reproduce the observed loops of Mercury's retrograde motion. Thus Copernicus was forced to add two small circles of equal size, 'stationed about the center of the greater epicycle, their axes being parallel to the axis of the deferent.' This innovation allowed Copernicus to explain the motion of Mercury in longitude. He goes on to say that 'Its motion in latitude is exactly like that of Venus, but always in the opposite direction.'

Copernicus was satisfied that he had explained all of the celestial motions with his heliocentric theory, in which the apparent irregularities in the orbits of the moon and the planets could be explained from the motions of the earth, three in number, to which his system of epicycles and eccentrics added additional circles. As he writes proudly in the last paragraph of the *Commentariolus*: 'Thus Mercury runs on seven circles in all; Venus on five; the earth on three; and round it the moon on four, finally Mars, Jupiter and Saturn on five each. Altogether, therefore, 34 circles suffice to explain the entire structure of the universe and the entire ballet of the planets.'

Despite its brevity, the *Commentariolus* contains all of the main ideas that Copernicus would later introduce in *De revolutionibus*, most notably the revelation that the sun and not the earth was the centre of the universe. Although he says in the beginning of the *Commentariolus* that he will leave mathematical demonstrations for his larger work, he describes the motions of the celestial bodies at some length, indicating that he has already worked out the details of his system, though he will alter parts of his planetary model in the final version of *De revolutionibus*.

Thus some three decades before the appearance of *De revolutionibus* Copernicus had fully developed his heliocentric theory. This involved a tremendous amount of work, which he would have done in the first decade after his return from Italy, during most of which he was extremely busy in the service of his uncle Lucas. It would then seem as if he left his uncle's service so as to return to the cathedral at Frauenburg, where he would have more time for his astronomical observations and calculations.

CHAPTER 7
THE LETTER AGAINST WERNER

Aside from his astronomical work Copernicus had been very busy during his canonical duties in the cathedral chapter. A record in the cathedral archives shows that in November 1516, at the chapter's annual meeting, he was elected 'Administrator of the Property Held in Common by the Venerable Chapter of Warmia for a term of three years beginning on St Martin's Day, the 11th of November.' Copernicus was elected to a second term as Administrator on 11 November 1520. He served until 31 May 1521, when he was replaced by Tiedemann Giese.

Meanwhile the Teutonic Knights had resumed their incursions from Prussia into Warmia. On 22 July 1516, Tiedemann Giese, writing on behalf of the Warmian chapter, petitioned King Sigismund to protect them from the Knights:

> For the past seven years the population of Warmia has endured bloody assaults and acts of robbery directed by the Teutonic Order. As recommended at the last meeting of the Estates of Royal Prussia, the chapter has begun to resist. Two weeks ago, when robbers attacked a citizen of Elblag and cut off his hands, we sent a small detachment into Teutonic Prussia, caught one of the robbers, a nobleman, and retrieved his booty. He was taken into custody along with his horses and weapons. The Grand Master of the Teutonic Order has demanded their return. Also the robbers have intensified their activities. The chapter begs the king to protect them from their enemies.

On 1 January 1520 the Teutonic Knights, with an army of 5,000 troops and cavalry, captured the town of Braunsberg, six miles east of Frauenburg. On 23 January they attacked Frauenburg and set fire to the town, destroying most of the houses, including those of Copernicus and the other canons. The canons were forced to seek shelter elsewhere, Copernicus and another colleague going to Allenstein. The cathedral was saved by the belated arrival of Polish troops sent earlier from Elbing.

The Knights invaded Warmia again that autumn, and on 19 October they invested Heilsberg Castle, forcing the cathedral chapter to take refuge in Allenstein. Then on 15 November the Knights captured the walled town of Gutstadt (Dobre Missto), 16 miles north of Allenstein. The following day Copernicus wrote to King Sigismund from Allenstein asking that the garrison of 100 troops be reinforced, noting that the 'canons want to act nobly and honestly as faithful subjects of the king and they are even prepared to die.'

Copernicus then sent a convoy of wagons to Elblag, asking his fellow canons there to send him food and clothing, as well as 20 small cannons to help defend Allenstein against the Knights. The other canons had by then fled from Allenstein, leaving Copernicus in charge of its defence. At the end of November a detachment of 100 Polish troops arrived, forcing the Knights to lift the siege of Allenstein, although they continued to occupy Gutstadt.

The Knights attacked Allenstein again on 16 January 1521, and ten days later they broke through one of the town gates before they were driven back by the garrison. Then in mid-February the Knights lifted the siege, after which their Grand Master, Duke Albrecht of Prussia, began negotiating for a settlement. A four-year truce was signed in Thorn on 9 April that ended hostilities, though a large part of Warmia remained under the occupation of the Teutonic Knights.

Copernicus was part of the delegation that represented Warmia at the peace negotiations. On 31 May he was appointed as *Warmia commissarius*, one of his duties being to superintend the reconstruction of Frauenburg. The rebuilding of Frauenburg proceeded rapidly, and by the end of the year Copernicus returned there to live.

During the two terms that Copernicus served as Chapter Administrator he had a wide variety of duties for which his university education had hardly prepared him; nonetheless he carried them out with distinction.

The Letter Against Werner

The chapter's laws required that the 'elected administrator shall swear that he will present to the chapter or its deputies in its entirety the whole income from the districts of Melsac (Pieniezno) and Allenstein.' This income was collected from the peasants bound to the land owned by the Chapter and divided into parcels.

After Copernicus was elected to this post he moved temporarily to Allenstein, living in the old Prussian castle. One of the entries he made in the surviving ledgers he kept gives some idea of his administrative activities, as well as of the conditions of peasant life at the time. This concerns a visit he made to the village of Jonikendorf (Jonkowo) on 10 December 1516.

> Merten Caseler took possession of 3 parcels [vacant] because Joachim was hanged on account of theft. They were not sown last year. I canceled the payment for this year, and he will pay next year and thereafter [...] He got 1 cow, 1 heifer, an ox and a sickle and, as for grains, a sack of oats and barley for the sowing omitted by his predecessor [...] In addition, I promised him 2 horses. The overseer was his guarantor for 4 years.

Another function of the chapter administrator was to approve certain financial transactions, in which all or part of the rent of a farm cultivated by a tenant went to the owner of the property. Four of these transactions were approved by Copernicus while he was Chapter Administrator, three of them at Allenstein in 1518 and the fourth at Frauenburg the following year.

Early in Copernicus' second term as administrator he reorganized the chapter's inventory of documents and legal papers. Since 1502 these had been kept in Allenstein because the castle there was more heavily fortified than Frauenburg. The entries in Copernicus' inventory comprise more than 161 documents listed in 17 categories, each of which was identified by a letter of the alphabet and kept in a separate drawer of a chest in the Chapter's treasury.

Drawer A held four documents concerning the establishment of the diocese of Warmia. As listed in Copernicus' inventory, these comprised the following historical documents:

> 2 Bulls of [Pope] Innocent VI concerning the boundaries of the regions of Prussia; 2 Golden Bulls of [Emperor] Charles IV concerning the delimitation of the regions of Prussia and the confirmation of the privileges

of the diocese of Warmia; Documents of the said Charles concerning the renewal of the privileges of the diocese of Warmia; Document of Anselm, the first bishop of Warmia, [concerning] the construction of a cathedral and the division [of the Prussian lands] into dioceses.

His duties as administrator of the chapter's property made Copernicus aware of a serious problem that had developed in the coinage of Poland. It seems that the Teutonic Knights were minting debased coins, using less silver and more copper than in the coinage issued by the King Sigismund. This led him to compose a brief essay in Latin on currency, completed on 15 August 1517 and entitled *Meditata* (Meditations), copies of which he gave to a few friends. The essay soon came to the attention of the West Prussian Estates, and at their request Copernicus prepared a German version of *Meditata*.

In the essay Copernicus recommended that the king controlled the minting of coinage and instituted reforms that would guarantee that all coins had their stated value. The basic idea that he put forward is that 'bad money drives out good money,' now known as Gresham's law. This is named after the English financier Sir Thomas Gresham (1519–79), who proposed it a quarter of a century after Copernicus.

Copernicus was asked to present his views at a meeting of the West Prussian Estates held in March 1522. Some three years later Copernicus composed a final version of the essay in Latin, entitled *Monetae cudendae ratio* (Essay on the Coinage of Money). The treatise was published posthumously in 1592, and after its rediscovery at the beginning of the nineteenth century it was reissued at Warsaw in 1816.

Copernicus begins the treatise by pointing out that debasement of currency is one of the principal causes for the downfall of countries, 'For it undermines states, not by a single attack, all at once, but gradually and in a certain covert manner.' He then presents his arguments for a sound currency:

> But maybe someone will argue that cheap money is more convenient for human needs, forsooth, by alleviating the poverty of people, lowering the price of food, and facilitating the supply of all the necessities of human life, whereas sound money makes everything dearer, while burdening tenants and payers of an annual rental more heavily than usual [...] But if they will have regard for the common good, they will surely be unable to deny that sound money benefits not only the state

but themselves and every class of people, whereas debased coinage is harmful.

Copernicus concludes his treatise with a number of recommendations, one being that the Prussian coinage should be adjusted to the Polish, another stating that, 'if possible, only one place should be designated for a mint.' He foresaw that the latter recommendation would be rejected by Duke Albrecht of Prussia, and so he compromised by saying that 'if, however, this could not be done because of the opposition of the Duke of Prussia on the grounds that he wants to have his own mint, let two places be designated at the most, one in his Royal Majesty's territory and the other in the Duke's domain.'

On 17 July 1526 King Sigismund met in Danzig with the West Prussian Estates, issuing a declaration agreeing with several recommendations that had been made by Copernicus. This was the first of a number of meetings on currency reform that were held in the years that followed. One of the most important of these was held at Elblag on 16 March 1528. One week earlier, on 9 March, Bishop Maurice Ferber wrote a letter to the cathedral chapter of Warmia, asking that Copernicus be selected to accompany him to the meeting. Copernicus attended several subsequent meetings on currency reform, the last in October 1530. After that the meetings dragged on for a time before they were finally abandoned.

Despite all these distractions Copernicus continued with his researches on astronomy. He gave no public mention of his astronomical views until 1524, when he responded to a letter he had received from his friend Bernard Wapowski, who had been his fellow student at the University of Krakow and was now a canon of the cathedral at Krakow and secretary to the King of Poland. Wapowski has asked Copernicus to give his opinion on a treatise published in Nuremberg in 1522 by Johannes Werner (1468–1522), a parish priest who was noted as a mathematician, astronomer, geographer and instrument maker. Werner's treatise, published around the time of his death along with a work on conic sections, is entitled *De motu octavae sphaerae*, known in English as *The Motion of the Eighth Sphere.*

The eighth sphere of Werner's title is that of the fixed stars, which in Aristotelian cosmology is the outermost of the celestial globes, enclosing those of Saturn, Jupiter, Mars, Venus, the sun and the moon, with the immobile earth at their centre. The eighth or stellar sphere, in the original Aristotelian model, rotated once a day about the celestial poles, in that way producing the apparent nightly rotation of the stars. In order to account for the precession of the equinoxes discovered by Hipparchus, Ptolemy attributed to the stellar sphere a slow rotation about the axis of the ecliptic in addition to its daily rotation. The Aristotelian model had each celestial sphere rotating on its own. In order to conform to this rule, Ptolemy, in his *Planetary Hypotheses*, postulated another sphere outside that of the stars which moved only with the diurnal rotation, which it transmitted to the stellar sphere, whose basic motion was now just the slow rotation required to account for precession.

During the Middle Ages the spurious trepidation theory described by Theon of Alexandria was revived when an Arabic work, now lost, on the motion of the eighth sphere was translated into Latin as *De motu octavae sphaerae*. This work is generally attributed to Thabit ibn Qurra, although the attribution has in recent years been questioned. The trepidation theory says that the precession of the equinoxes, rather than being a constant periodic motion, as stated by Hipparchus and Ptolemy, instead reverses direction every 640 years, with the equinoxes moving back and forth along the ecliptic over a range of eight degrees. The trepidation theory was used in combining the *Alfonsine Tables*, which took the period of precession to be 49,000 years, as compared with the modern value of approximately 25,800 years.

Copernicus discussed the precession of the equinoxes in *De revolutionibus*, III, 1. There, while rejecting the notion that the intersection points of the equinoxes and ecliptic oscillated, he proposed that the rate of precession is variable, and he sought to determine its period and mean value. He did this by measuring the longitudes of the stars Spica and Regula, which he compared to values recorded by Arabic and Greek astronomers, including Hipparchus and Ptolemy. He used this information to formulate a geometric theory to explain both the supposed change in the rate of precession and the slow decease in the obliquity of the ecliptic that had been recorded since antiquity.

Werner wrote about both the trepidation theory and the variation of the obliquity in his thesis, but his ideas conflicted with those of Copernicus, who was thus led to respond in detail to Wapowski's letter.

Copernicus answered in letter, dated 3 June 1524, severely criticizing Werner's views. Handwritten copies of his response, which in English came to be called *The Letter Against Werner*, circulated privately throughout Europe, and from one of them preserved in Berlin the first printed edition was made, which was included in an edition of *De revolutionibus* published in Warsaw in 1854 by Jan Baranowki. A second copy was discovered in Vienna by Maximilian Curze, who collated both manuscripts and published a critical text in 1878. The first English translation was made by Edward Rosen and published in 1939 in his *Three Copernican Treatises*.

Copernicus begins his response to Wapowski with a hint of the criticism to come:

> Some time ago, my dear Bernard, you sent me a little treatise on *The Motion of the Eighth Sphere* written by John Werner of Nuremberg. Your Reverence stated that the work was widely praised and asked me to give you my opinion of it. Had it been really possible for me to praise it with any degree of sincerity, I should have replied with a corresponding degree of pleasure. But I may commend the author's zeal and effort.

He goes on say that he did not want to give a negative appraisal of Werner's views without pointing out how they were in error, so as to clarify the subject under discussion: 'Consequently, lest I seem to condemn the man gratuitously, I shall attempt to show as clearly as possible in what respect he errs regarding the motion of the sphere of the fixed stars and maintains an unsound position. Perhaps my position may even contribute not a little to the formation of a better understanding of the subject.'

Copernicus begins his critique by pointing out a serious error Werner had made in the calculation of time intervals, since this was crucial in determining the motion of the eighth sphere. Werner, in Proposition 4 of his thesis, said that Ptolemy's star catalogue compiled in the second year of the reign of Emperor Antoninus Pius was done 'in the 150th year of the incarnation of our Lord.' Copernicus corrects this date by 11 years, in which he uses the Greek chronology beginning the first year in the reign of the Egyptian pharaoh Nabonassar (r.747–33 BC).

He then highlighted another serious mistake that had been made by Werner in one of his hypotheses: Copernicus says that 'The hypothesis in which he expresses his belief that during the four hundred years before Ptolemy the fixed stars moved with equal motion only involves a second error no less important than the first.' He is referring to the statement that Werner made in Proposition 8 of his thesis: 'Therefore it is clear that the fixed stars moved only with equal motion, and lacked unequal motion; or if they had any unequal motion, it was very small and almost imperceptible.' Then in the corollary to this proposition he states that 'Hence it is clear that the fixed stars in their equal motion complete one revolution in 36,000 years.'

Here Werner was referring to the time taken for the equinoxes to complete one full precession, which he said is constant. This ran counter to Copernicus' theory that the rate of precession is not constant but variable.

Werner concluded that the motion of the eighth sphere, which he said had been constant in the 400 years before Ptolemy, had been more rapid in the period from Ptolemy up to the time of King Alfonso, for whom the *Alfonsine Tables* had been compiled. Copernicus refutes this idea with some sarcasm by pointing out that Werner had a confused notion of 'uniform' motion and would not have drawn this conclusion if he had consulted his own astronomical tables.

Copernicus then repudiates another of Werner's conclusions, where he implied that the Greek astronomer Timocharis, a contemporary of Aristarchus, had erred in his observations of the coordinates of two stars, as recorded by Ptolemy, since they differed from the values that Werner had computed in his own tables. Copernicus also indicates that Werner's error of 11 years in dating Ptolemy's star-catalogue added to the error he made in computing of the two stars in question.

Copernicus' theory that the rate of precession is not constant but variable was a view that he seems to have had long before he stated it in *De revolutionibus*. As Edward Rosen explained, this can be inferred from a close reading of the two passages in *De revolutionibus* concerning the number of celestial spheres.

The first passage is in Book I, 11, where Copernicus discusses the third of the three motions of the earth, namely the precession of the equinoxes:

> Now we said that the yearly revolutions of the centre and of the declination were nearly equal, because if they were exactly so, then the points of equinox and solstice and the obliquity of the ecliptic in relation to the sphere of the fixed stars could not change at all. But as the difference is slight, it is not revealed as it increased with time: as a matter of

fact, from the time of Ptolemy to ours there has been a precession of the equinoxes and solstices of about 21°. For that reason some have believed that the sphere of the fixed stars was moving, and so they chose a ninth higher sphere. And when that was not enough, the moderns added a tenth, but without attaining the end which we hope we will attain by means of the movements of the Earth. We shall use these movements as a principle and a hypothesis in demonstrating other things.

The second passage is in Book III, 1, where, in discussing the precession of the solstices he says, 'Moreover, there is an additional surprise of nature in that the obliquity of the ecliptic does not appear to us so great as before Ptolemy – as we said before: "For the sake of" a cause for these facts some have thought up a ninth sphere, and others a tenth: they thought that these facts could be explained through these spheres but they were unable to produce what they promised. Already an eleventh sphere has begun to see the light of day, and in talking of the Earth we shall easily prove that this number of spheres is superfluous.'

Copernicus' reference to the tenth sphere comes from two statements by Werner, in Propositions 11 and 13, the first of which states:

> The apparent or unequal motion of the sphere of the fixed stars or of the eighth sphere is caused by the circumstances that the first points of Cancer and Capricornus of the ecliptic of the ninth sphere revolve on small circles. This revolution is called by Thabit and the Alfonsine Tables the forward and backward or trepidation of the eighth sphere. This trepidation proceeds sometimes in the order of the signs, sometimes in the contrary order. Hence the motion of the fixed stars is sometimes slow and sometimes rapid. It is clear, moreover, that the motion of the fixed stars is composed of the equal motion of the eighth sphere, and the trepidation or forward motion of the ninth sphere on the small circles.

Then in Proposition 13 Werner declares: 'Therefore it is clear that the first points of Cancer and Capricornus of the ninth sphere were, about Ptolemy's time, near the aforesaid intersection of the small circles with the ecliptic of the tenth sphere.'

Rosen supposed that Werner added an eleventh sphere for the diurnal rotation of the entire universe, the *primum mobile*, or prime mover, as it is shown in a representation of Aristotle's cosmology in the *Cosmographia* of Petrus Apianus, published in Antwerp in 1539. Rosen thus concluded that it was Werner who was responsible for introducing the additional sphere which led Copernicus to state, in *De revolutionibus* III, 1, that 'Already an eleventh sphere has begun to see the light of day.'

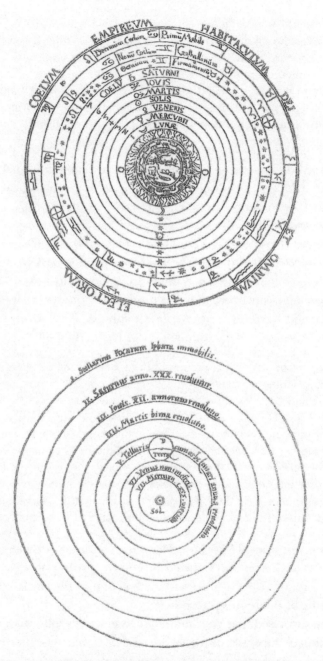

Figure 4 Aristotle's geocentric theory, Peter Apian, *Cosmographica*, 1539 (above); the Copernican heliocentric theory, *De revolutionibus*, 1543 (below).

This led Rosen to assume that Copernicus wrote section I, xi of *De revolutionibus*, which refers only to a ten-sphere cosmos, before he saw Werner's treatise:

> Hence he saw the passage in I, 11, before he saw Werner's *Motion of The Eighth Sphere*. For had Werner's eleventh sphere been known to Copernicus when he was writing I, 11, he would surely have fortified his position there by mentioning an eleventh sphere. Instead, in I, 11, he says: 'more recent writers now add on a tenth sphere.' Clearly, Copernicus wrote I, 11, before he read Werner, and he wrote III, 1, after he read Werner. He dated his *Letter Against Werner* 3 June 1524. Hence he wrote *Revolutions*, I, 11, before that date, which is of course much earlier than 1530.

Rosen then gave his reasons for thinking that Copernicus began writing *De revolutionibus* well before 1530, probably no later than 1515. He presented his arguments in a paper entitled 'When did Copernicus Write the Revolutions?', which is included in his *Copernicus and His Successors*.

There he said 'But Copernicus was a taciturn man, who talked very little about himself and his work. For instance, he does not say precisely when he began to write the "Revolutions". Nevertheless he does provide a clue.' Rosen went on to say that Copernicus gives this clue in the *Commentariolus*:

> 'Here', he explains, 'for the sake of brevity I have sought it desirable to omit the mathematical demonstrations intended for my larger work'. The only 'larger work' ever written by Copernicus is his 'Revolutions', in which he inserted the mathematical demonstrations omitted from his 'Commentariolus'. Evidently he was already planning the 'Revolutions', or perhaps he had already begun to write that 'larger work', when he made this reference to it in his 'Commentariolus'.

One of Rosen's arguments refers to the letter, described earlier, that Copernicus wrote to Paul of Middleburg in connection with the papal project on calendar reform:

> Now Copernicus replied to Paul of Middleburg in or about 1515. What he found 'particularly in 1515 AD' [...] made him modify two of the main tenets upheld in his 'Commentariolus': the invariability of the

earth-sun (a) eccentricity and (b) apsidal direction. Copernicus' dating of the composition of *Revolutions*, III, 16, is not as precise as we might wish. But when he there speaks of 'the more than ten years since I have devoted my attention to investigating these topics, and particularly in 1515 AD'[...] he surely implies that he wrote *Revolutions*, III, 16, before 1530.

Rosen reinforced his argument with reference to observations that Copernicus discusses in *De revolutionibus*. In *De revolutionibus*, I, 10, Copernicus mentions 'that, according to Ptolemy, there are 38 earth-radii to the moon's perigee.' But he immediately adds that 'according to a more accurate determination there are more than 49, as will be made clear below.' Then in *De revolutionbus* IV, 17, 22, 24, he fixes the distance of the moon's perigee at more than 52 earth-radii. The inconsistency with the earlier result was removed in the first edition of *De revolutionibus*, but it remains in folio 8 of Copernicus' autograph, the original manuscript. Rosen concluded that when Copernicus wrote Book I, 10, with the value of 'more than 49 earth-radii,' he did not yet have the higher value he quotes in the later sections. Rosen went on from this to support his conclusion that the original version of *De revolutionibus*, I, 10, was written well before 1530.

> The higher value was derived from the calculations discussed in the *Revolutions*, IV, 16. These two observations are dated 27 September 1522 and 7 August 1524. Therefore Copernicus wrote I, 10 (and folio 8 in his autograph) before he drew his conclusions from these observations of 1522 and 1524. Folio 8 was written on paper C, one of the four batches of paper used by Copernicus for his autograph. The observations of 1522 and 1524 are reported on folios 126 and 127, which belong with paper D, another of Copernicus' four batches. He generally used paper C before paper D, as he did here, but the opposite also occurs. In any case, he wrote folio 8 and *Revolutions*, I, 10, long before 1530.

Copernicus concludes the *Letter Against Werner* with a reflection upon his own views concerning the motion of the eighth sphere, which he says he intends to give elsewhere, obviously referring to the work that would be called *De revolutionibus*:

> So much for the motion in longitude of the eighth sphere. From the foregoing remarks it can easily be inferred what we must think about

the motion in declination, which our author has complicated with two trepidations, as he calls them, piling a second one upon the first [in Proposition 18]. But since the foundation has now been destroyed, of necessity the superstructure collapses, being weak and inconclusive. What finally is my own opinion concerning the motion of the sphere of the fixed stars? Since I intend to set forth my views elsewhere, I have thought it unnecessary and improper to extend this communication further. For it is enough to satisfy your desire to have my judgment of this work, as you requested. May your reverence be of sound health and good fortune.

But nearly two decades would pass before the world would learn the revolutionary views of Copernicus. Meanwhile he would continue his observations and calculations while carrying out his duties as a canon in Frauenburg cathedral, far away from the cultural centres of Renaissance Europe.

CHAPTER 8
THE FRAUENBURG WENCHES

While he was engaged in his astronomical researches Copernicus continued to be a full-time administrator looking after the properties of the cathedral chapter, which owned almost half of the land in Warmia, with some of the places under his supervision requiring two days of travel to visit.

In 1531 Copernicus and Tiedemann Giese were appointed Guardians of the Chapter's Counting Table. They were responsible for collecting the money owed to the chapter by purchasers who had bought some of its land and paid in instalments. Other financial responsibilities included looking after the chapter's investments, including a mill which was leased to provide a steady dividend. These investments were financed from special funds belonging to the chapter, and the Guardians collected the income from these properties. They also managed the various bequests received by the chapter.

On 30 April of the following year they submitted to the chapter three related documents for approval, the first and third written by Giese and the second by Copernicus. The one submitted by Copernicus concerned a mill in Haselau bought by the chapter from George of Baysen. His report is known in English as the *Allenstein Bread Tariff*, though it does not actually concern a tax. Rather, it is an analysis of what is required to keep the cost of a loaf of bread equal to one penny, which Copernicus believed to be a 'just price.' The guidelines recommended by Copernicus are remarkably detailed, in the measurements as well as the technical and economic analysis involved:

> From one sack of both grains [wheat and rye] after a careful weighing has been conducted and [the weight] of the bread basket has been subtracted, about 67 pounds of bread are produced. But since the darnel and tares are usually separated from the grain before it is ground in order that the bread may come out cleaner and purer, it was agreed heretofore to subtract one pound for such cleansings, so that at least 66 pounds of bread result from one sack. Furthermore, the ordinary expenditures are 6 shillings, 4 pence, namely the baker's usual 4-shilling wage, 1 shilling for transportation, 1 shilling for salt and yeast, 4 pence for sifting. But the bran and chaff are enough to cover the baking expenses, provided that 8 sacks of bran invariably sell for 6 shillings. Hence, the same constant ratio for the price of grain to the production of loaves prevails. Thus, for example, when grain is bought for 33 shillings [a sack], 6 penny loaves will weigh 2 pounds. But when the price is 22 [shillings a sack], 6 loaves ought to weigh 3 pounds, and so on, as in the Table below [not shown] beginning with 9 [shillings] and increasing by 3 [shillings].

An undated document in the chapter archives records that Copernicus supervised the maintenance of the clock of Frauenberg Cathedral, whose repair cost 3 ducats. Another document, dated 1537, records that Copernicus was appointed to the office of Protector of the execution of testaments and fortifications.

Copernicus also continued to serve as personal physician to the Bishop of Warmia, a post which he had held since 1507. After the death of his uncle Lucas Watzenrode the next bishop was Fabian von Lossainen, who was elected on 5 April 1512. On 8 May 1519 Bishop Fabian wrote to canon Tiedemann Giese giving his condolence for the death of Giese's two sisters, who had died of the plague. He also asked Giese to request the help of 'Doctor Nicholas and other physicians' in trying to stop the plague from spreading farther.

Bishop Fabian died on 30 January 1523 and was succeeded by Maurice Ferber, who was elected on 16 April of that year. On 10 July 1529 Bishop Ferber wrote to canon Alexander Scultetus, a younger colleague of Copernicus', thanking him for sending him a map and description of Livonia, the region north of Lithuania. At the end of the letter he said, 'Besides, you should talk about our project and work in common intelligence with Doctor Nicolaus, so that we can have a map and a description of the lands of Prussia.'

Copernicus had a very high regard for Scultetus, whom he described as 'the only one outstanding in every respect among all the officials and canons of our Cathedral Chapter of Warmia'. Scultetus had the same high opinion of Copernicus, whom he refers to as 'Nicholas Copernicus, canon of Fromberk, astronomer and mathematician,' in his *Chronology or Annals of Nearly All the Kings, Princes and Potentates of the World to the Year 1545*, published at Rome in 1546.

Scultetus had written a treatise entitled *Catalogus rerum Prutenicarium et praesertim Warmiensium* (Catalogue of Things from Prussia and especially from Warmia), now lost. The map and description of Livonia mentioned by Bishop Ferber were probably from this work. Copernicus is thought to have collaborated with Scultetus on this treatise, since he is known to have worked on cartography. As evidence of this, on 17 February 1519 Bishop Ferber wrote to Tiedemann Giese about a map on which Copernicus had worked. The treatise probably served as the basis for a map of Prussia drawn by the young cartographer, Heinrich Zell, who would later visit Copernicus in Frauenburg.

Meanwhile the Protestant Reformation had brought about a seismic change in the world of Copernicus and his contemporaries in Prussia and Poland, beginning with Martin Luther's posting of his *Ninety-five Theses* at Wittenberg in 1517. Nuremberg officially adopted Lutheranism in 1524, as did a number of other cities as well as principalities in the German territories. Sweden adopted Lutheranism in 1529, and that same year Frederick I of Denmark converted, forcing his country to follow him.

On 28 March 1523 Luther wrote to the Teutonic Knights telling them to break their vows of celibacy and take wives. The first to renounce his vows was Georg von Polenz, Bishop of Samland as well as Regent and Grand Chancellor of Prussia, who on Christmas Day, 1523, preached a sermon in which he urged the Teutonic Knights to follow his example. He persuaded the Grand Master Albrecht of Hohenzollern-Ansbach that he would be unable to become a Polish vassal as long as he was a Roman Catholic, but as a Protestant Duke he could do as he pleased. Albrecht had been converted to Lutheranism by the young theologian Andreas Osiander, Nuremberg's first Protestant minister, who would later play a role in the publication of Copernicus'

De revolutionibus. The Grand Master was anxious to begin a dynasty, and so, by July 1524, he decided to renounce his vows, marry and convert Prussia to a secular principality under his own rule, though the conversion did not become official for a few years.

The four-year truce between the Teutonic Knights and the Polish Kingdom agreed to in the Peace of Thorn expired on 8 April 1525. Albrecht knew that he could not afford renewed hostilities, and so on 8 April 1525 he agreed to a resumption of the truce in the Treaty of Krakow.

Two days later Albrecht knelt before Sigismund I and swore fealty to him, whereupon the king made him Duke of Prussia as his hereditary possession and a fief of the Polish Crown. Then on 27 May of that year, before the assembled representatives of the Teutonic Knights, Duke Albrecht dissolved their religious order. But many of the knights did not accept this, and the Order elected a new Grand Master, Walter von Cronberg, who received Prussia as a fief at the Diet of Augsberg.

Albrecht officially announced that he had become a Lutheran in August 1525, and four months later he issued a church ordinance that converted his duchy to Lutheranism. The following year he married Princess Dorothea, the daughter of King Frederick I of Denmark, who in the next 13 years bore him four daughters and two sons, both of whom died in infancy. After the death of Dorothea he married Anna Maria, daughter of Eric I, Duke of Brunswick-Lüneburg, who bore him two children, first a daughter and then a son, Albrecht Friedrich, who succeeded as Duke of Prussia when Albrecht died on 20 March 1568.

Meanwhile Lutheranism had begun to penetrate the bishopric of Warmia. Riots in Danzig in 1525 forced the city council to adopt Lutheranism. King Sigismund responded by mustering a force of 8,000 soldiers, and in the summer of 1526 he personally led them in a campaign that drove the Lutheran leaders out of Danzig. In September of that year Bishop Ferber issued an edict expelling all Lutherans from Warmia, as well as ordering the destruction of all pro-Lutheran publications. Nevertheless, Lutheranism revived five years later in Elblag, some 20 miles southwest of Frauenburg, where Dutch Lutherans had recently settled and begun proselytizing. On Shrove Tuesday 1531 the Lutherans in Elblag put on a play in which the pope and the high-ranking Catholic clergy were lampooned. This outraged Bishop Ferber who took vigorous measures to have those responsible punished.

The Frauenburg Wenches

All of this seems to have so alarmed the bishop that he began to take steps to bring his own clergy into line. All the canons of the Frauenburg chapter had taken 'first orders,' which included a vow of celibacy, but only one of them had gone on to take the 'higher orders' necessary to qualify as a priest and celebrate mass. Ferber complained of this in a letter he sent on 4 February 1531 to Copernicus and the other canons who had not taken higher orders, threatening to cut off their income unless they complied, but there is no indication that he ever carried out his threat.

Later that year Copernicus received at least two more letters of complaint from Ferber, the second of which he replied to on 27 July 1531. Here the bishop refers to an unnamed woman who had been Copernicus' housekeeper before she married a man from whom she later separated, after which she went to work for a woman in Elblag. Copernicus begins the letter by addressing Ferber as 'My lord, Most Reverend Father in Christ, my noble lord':

> With due expression of respect and deference, I have received your most Reverend Lordship's letter. Again you have deigned to write to me with your own hand, conveying an admonition at the outset. In this regard I must most humbly ask your Most Reverend Lordship not to overlook the fact that the woman about whom your Most Reverend Lordship writes to me was given in marriage through no plan or action of mine. But this is what happened. Considering that she had once been my faithful servant, with all my energy and zeal I endeavoured to persuade them to remain with each other as respectable spouses. I would venture to call upon God as my witness in this matter, and they would both admit it if they were interrogated. But she complained that her husband was impotent, a condition which he acknowledged in court as well as outside. Hence my efforts were in vain [...] However, with reference to the matter, I will admit to your Lordship that when she was recently passing through here from the Krolewiec fair with the woman from Elblag who employs her, she remained in my house until the next day. But since I realize the bad opinion of me arising therefrom, I shall so order my affairs that nobody will have any proper pretext to suspect evil of me thereafter, especially on account of your Most Reverend Lordship's admonition and exhortation. I want to obey you gladly in all matters, and I should obey you out of a desire that my services may always be acceptable.

The woman in question was Anna Schilling, who had worked for Copernicus as his housekeeper. (She is referred to in the sources as a *focaria*, which can mean either 'housekeeper' or 'concubine'.) She is believed to be the daughter of a Dutchman, Arend van der Shelling, who had changed his name to Schilling when he moved to Danzig and married a distant relative of Copernicus. During the years 1529–36 Schilling and Copernicus acted as the legal guardians of a group of orphaned children, and it was probably this connection that brought Nicolaus and Anna together. Absolutely nothing is known of her marriage, except the fact that she took action to have it annulled on the grounds of her husband's impotence. Anna had her own house in Frauenburg, or at least she did before she went to work in Elblag, which would have been after Copernicus had felt it necessary to terminate her employment because of Ferber's criticism. But, as events showed, this was not to be the end of their relationship.

At the end of 1531 Bishop Ferber was stricken with a severe illness, which led him to summon Copernicus to come and treat him. On 29 December the bishop wrote to Laurence Wille, personal physician of the Duke of Prussia, asking for his help and giving Copernicus' description of his symptoms. Then on 10 January 1532 he wrote to Johannes Benedict Solfa, personal physician of the King of Poland, asking him to send more medicines in case of a recurrence of the illness that 'Doctor Nicolaus and Doctor Wille are successfully fighting.'

Meanwhile word of Copernicus' accomplishments in astronomy seem to have spread slowly in learned circles in Europe, probably through copies of the *Commentariolus* that had been circulated. There is also mention of an almanac based on new astronomical tables that Copernicus seems to have compiled. The only direct reference to this almanac is by his old friend Bernard Wapowski, who last saw Copernicus early in the autumn of 1535. Wapowski sent the almanac to Baron Sigismund von Halberstein, a diplomat in the service of Emperor Maximilian, in a letter dated 15 October 1535, where he gives a garbled description of Copernicus' revolutionary theory:

> I am sending you something new and long awaited by learned men: an almanac with the most trustworthy and correct motions of the planets, an almanac departing from common almanacs, and computed from new tables established by Sir Nicolaus Copernicus, canon in Warmia. He notes that Mercury goes away by half a constellation and also that the configurations of planets differ by several weeks from those in the

old almanacs. Sir Nicolaus is a very great mathematician, he claims that in order to verify the motions of the planets, one must admit the earth to be moving; he professes this opinion for many years now and assures that the motion of the earth is not perceivable [...] I would like that this thing be widespread, most of all among the men who know the celestial bodies and who establish almanacs in Germany, in order for them to make better ones and acknowledge having been in falsehood and having given erroneous tables too.

Wapowski died on 21 November 1535, and no record of any reply to his letter exists. Despite the lack of further allusion to Copernicus' almanac, there is a reference to planetary tables that he seems to have compiled some time before the publication of *De revolutionibus*, although these may have been the tables that were included in that work. This reference is in a singular letter, dated 1 November 1536, which Copernicus received from Cardinal Nicolaus Schönberg, the Papal Nuncio in Poland, who seems to have heard of his revolutionary theory.

> Some years ago, word reached me concerning your proficiency of which everybody constantly spoke. At that time, I began to have a very high regard for you, and also to congratulate our contemporaries among whom you enjoyed such great prestige. For I had learned that you had not merely mastered the discoveries of the ancient astronomers uncommonly well but had also formulated a new cosmology. In it you maintain that the earth moves, that the sun occupies the lowest, and thus the central place, in the universe [...] I have also learned that you have written an exposition of this whole system of astronomy, and have computed the planetary motions and set them down in tables, to the greatest admiration of all. Therefore with the utmost earnestness I entreat you, most learned sir, unless I inconvenience you, to communicate this discovery to scholars. I have instructed Theodoric of Reden to have everything copied in your quarters at my expense and dispatched to me. If you gratify my desire in this matter, you will see that you are dealing with a man who is zealous with your reputation and eager to do justice to so fine a talent.

There is no record of Copernicus having responded to Schönberg, who died in August or September 1537. Copernicus preserved the letter and had it printed as a headnote to *De revolutionbus*, just before the preface.

There he writes of the friends who encouraged him to overcome his reluctance and publish his work:

> First among them was Nicholas Schönberg, Cardinal of Capua, a man distinguished in all branches of learning.

Schönberg seems to have been informed of the new theory by his secretary, Johann Albrecht Widmanstetter, who may have learned about it himself from one of the canons at Frauenburg or from reading a copy of the *Commentariolus*. Widmanstetter had previously served as a papal secretary under Clement VII, and had entered Schönberg's service after the pope's death on 5 September 1534. During the summer of 1533 he had given a lecture in the Vatican gardens entitled *'Copernicana de motu terra sontoniam explicani'* (An explanation of Copernicus' opinion about the earth's motion), in front of Pope Clement VII, the two cardinals, Francesco Ursini and Johann Salviati, the bishop of Viterbo, Johann Petri, and the physician Mathaus Curtius. The pope rewarded Widmanstetter with a rare Greek codex, *De sensu et Sensibilibus*, written by Alexander of Aphrodisias, a sixth-century AD commentator on Aristotle. The manuscript is preserved in the State Library of Munich, its provenance established by 11 handwritten lines in the hand of Widmanstetter.

Another reference, though vague, to the growing reputation of Copernicus, comes in a letter sent from Nuremberg on 8 April 1535 by Doctor Johann Apelt to Duke Albrecht, enclosing a horoscope worked out by Joachim Cammermeister (Cammarius). Apelt advised the duke to find somebody who could explain the horoscope to him, possibly, as a last resort, *'ein alter thumherr zur Fraunburg'* (an old canon from Frauenburg).

Bishop Ferber died at Heilsberg on 1 July 1537, and on that same day the cathedral chapter in Frauenburg issued an official announcement concerning the legal matters involved:

> The Chapter of Warmia notifies through a mandate to the Bishop's officeholder that, because of the death of Maurice Ferber, Bishop of Warmia, on July 1, 1537, it has taken over complete authority for the Warmian lands until the election of his successor. It has delegated to Heilsberg as its plenipotentiary administrators and legal advisors Nicolaus Copernicus and Felix Reich, Canons of Warmia, to prepare

The Frauenburg Wenches

a list of the late Bishop's things, the things belonging to the Bishopric, and for them to take care of the affairs of the castles, towns, office-holders and subjects of the Bishop. The Chapter calls for the manifestation of obedience to its delegates.

Copernicus and Reich departed that same day in two coaches, along with their servants, spending the night at an inn before arriving at the castle in Heilsberg on 2 July. They were able to complete their work by 4 or 5 July, after which they loaded their coaches with the deceased bishop's remains and belongings and returned to Frauenburg. A few days later Ferber was interred in the cathedral along with other past bishops of Warmia, including Copernicus' uncle Lucas.

The chapter then made arrangements for an election to choose Ferber's successor. The list of candidates drawn by the royal chancellery included Johannes Dantiscus, Johannes Zimmerman, Achacy von Trenck and Henryk Snellenberg. When the list was made known in Frauenburg, on 21 August 1537, the canons proposed Copernicus rather than Snellenberg, 'whose designation would be regarded by everybody as ridiculous.' The amended list was then sent to King Sigismund for his approval.

Sigismund responded to the canons on 4 September approving the list, and he expressed his confidence that Johannes Dantiscus would be elected. That settled it, so far as the canons were concerned, and on 20 September they wrote back to the king informing him that Dantiscus had been unanimously elected.

Johannes Dantiscus (1485–1548), the son of a German brewer, was born Johann Flachsbinder, the Latin version of his name stemming from the fact that he was born in Danzig. He entered the University of Krakow in 1500, when he was 15, but interrupted his studies to join the army of King John I Albert, Jagiellon, fighting in two campaigns, first against the Tatars, and then against the ruler of Moldavia and Wallachia. Later he resumed his studies and graduated from the university in 1503, after which he joined the service of King Alexander, the last of the Jagiellonian dynasty, as a scribe in the royal chancellery. After a two-year pilgrimage to the Holy Land he returned to the royal service under the young Sigismund I, becoming the king's private secretary and soon afterwards his special envoy to the Prussian Estates. In 1515 he accompanied Sigismund to the convention of the

Holy Roman Empire in Vienna, where he was knighted by Emperor Maximilian I, who also made him poet laureate because of the Latin verses that he had published. He served as Sigismund's ambassador to the courts of Maximilian and his successor Charles V, who ennobled him and granted him an estate in Spain. His diplomatic missions took him to England, where he formed friendships with King Henry VIII and Cardinal Wolsey, and also to Sweden, where he became a friend of King Frederick I. He made a diplomatic visit to Wittenberg in 1523, meeting the two leaders of the reformation, Martin Luther and Philipp Melanchthon, whom he referred to in his correspondence as *gutten Gesellen* (good fellows).

Dantiscus also corresponded with Gemma Frisius (1508–55), the renowned mathematician, cartographer, philosopher and instrument maker. Through his correspondence with Dantiscus, Frisius first learned of Copernicus and his heliocentric theory, of which he was one of the first supporters.

During his diplomatic travels Dantiscus acquired a number of mistresses, only two of whom are known by name: Grinea, who was from Innsbruck, and Isabel Delgada, of Toledo. Isobel bore him two children, one of whom died in infancy; the other was a daughter named Juana Dantisca de Curis, born in 1527. Dantiscus always acknowledged that he was Juana's father, and regularly sent money to Isobel for her support.

Dantiscus left Spain soon after Juana's birth and returned to Poland, where he left the diplomatic service to pursue a career in the Church. He had had this in mind at least as early as 1514, when he applied for the canonry in the Warmia chapter that had been left vacant when Andreas Copernicus was excluded because of his leprosy, but Alexander Scultetus was chosen instead. He applied again in 1515 and yet again in 1528, but both times he lost out to candidates who had stronger support in Rome. He finally obtained a canonry in the Warmia chapter in 1529, and then on 4 May 1530 King Sigismund appointed him Bishop of Kulm, although he did not officially take up the position until September 1532. After Dantiscus became Bishop of Warmia, Tiedemann Giese replaced him as bishop of Kulm, a scenario that seems to have been worked out well in advance by King Sigismund, for Dantiscus had been his favourite since he first came to the throne.

The Frauenburg Wenches

Early in the autumn of 1532, Dantiscus wrote to Copernicus, inviting him and Felix Reich, the two most senior canons at Frauenburg, to visit him at his castle in Lubawa (Loebau), some 45 miles east of Kulm. Copernicus wrote back an apologetic letter to explain why he and Reich were unable to accept the invitation:

> I have received Your Most Reverend Lordship's letter and I understand well enough Your Lordship's grace and goodwill towards me, which he has condescended to extend not only to me but to other men of great excellence. It is, I believe, certainly to be attributed not to my merits, but to the well-known goodness of Your Rev. Lordship. Would that some time I should be able to deserve these things. I certainly rejoice more than can be said, to have found such a Lord and Patron. However, regarding Your Rev. Lordship's invitation to join him on the 20th of this month (and that I should most willing to do, having no little cause to visit so great a friend and patron), misfortune prevents me from doing so, as at that time certain business matters and necessary occasions compel both Master Felix and me to remain at this place.

There is no record of any further correspondence between the two until June 1536, when Copernicus wrote to Dantiscus, again regretfully declining an invitation to his palace, on this occasion to attend the wedding of a relative of the bishop. Otherwise, the only other known correspondence between them during Dantiscus' tenure as Bishop of Kulm was a letter from Copernicus giving him news from abroad.

The first recorded contact between them after Dantiscus took up his posts as Bishop of Warmia was in April 1538, when he became ill and summoned Copernicus to treat him in Heilsberg. In a letter to the canons at Frauenburg dated 11 April 1538, Dantiscus said that he wanted to keep Copernicus 'a few more days, until he felt better.' In a later letter he wrote 'I feel better, and you'll hear details from the doctor, our well-respected friend. His soft manners and talks, his pieces of advice, as much as I followed them, contributed to my healing.' In another letter Dantiscus noted that he had come to 'cherish [Copernicus] [...] as a brother.'

Near the end of July 1538, Copernicus and Reich were summoned to the bishop's palace at Heilsberg, from where they were to accompany Dantiscus on an official tour of the surrounding towns. There is no record of anything of note that transpired between the three men during this journey, when, particularly in the evenings, they would have had time to talk together with some degree of familiarity. Probably it was on this trip that Dantiscus learned of Copernicus' involvement with Anna Schilling, and the bishop may have revealed his own relationship with Isabel Delgada, with whom he was in correspondence concerning the coming marriage of their daughter Juana Dantisca.

In any event, Dantiscus warned Copernicus to end his relationship with Anna, whom he was continuing to see in his *curia*. Copernicus promised to end all contact with Anna, but Dantiscus seems to have learned that Copernicus continued to see her after his return to Frauenburg. And so in November 1538 he wrote to remind him of his promise. Copernicus responded, in a letter dated 2 December 1538, in which he tried to explain his difficult situation, for he needed a housekeeper for his *curia*, preferable a woman relative:

> My lord, Most Reverend Father in Christ, most gracious lord, to be heeded by me in everything: I acknowledge your quite fatherly, and more than fatherly admonition, which I have felt even in my innermost being. I have not in the least forgotten the earlier one, which your Most Reverend Lordship delivered in person and in general. Although I wanted to do what you advised, nonetheless it was not easy to find a proper female relative forthwith, and therefore I intended to terminate this matter by the time of the Easter holidays. Now, however, lest your Most Reverend Lordship suppose that I am looking for an excuse to procrastinate, I have shortened the period to a month, that is, to the Christmas holidays, since it could not be shorter, as your Most Reverend Lordship may realize. For as far as I can, I want to avoid offending good people, and still least your Most Reverend Lordship. To you, who have deserved my reverence, respect, and affection in the highest degree, I devote myself with all my faculties.

Copernicus wrote to Dantiscus again in mid-January 1539, assuring the bishop that Anna was no longer his housekeeper and that he had terminated his relationship with her: 'I have done what I neither would not or could have left undone, whereby I hope to have given satisfaction to your Rev. Lordship.'

But it turned out that Anna was still in Frauenburg, though perhaps in her own house if not in Copernicus' *curia*. Dantiscus may have suspected as much, and he wrote to Felix Reich, enclosing an official letter of disapproval of Copernicus' misconduct to be read before the entire chapter of canons. But Reich was a close friend of Copernicus', and so he delayed matters by sending the letter back to Dantiscus with suggestions for it to be reworded. He said he hoped that Copernicus would take the bishop's warning to heart so that it would not be necessary to admonish him, adding: 'He will be overcome with shame, I am afraid, if he learns that I am privy to this matter. Had I not been prevented by the insertion of certain little words, I could perhaps have read to him your Lordship's letter insofar as it touches on that business.'

By that time Dantiscus had become aware that two other canons in Frauenburg – Alexander Scultetus and Leonard Niederhoff – were involved with women. Scultetus lived openly with his mistress, with whom he had at least one child. Thus when Dantiscus wrote again to Reich he said that he would be sending him official writs of misconduct for Scultetus and Niederhoff as well as the amended letter to Copernicus. Reich responded by advising Dantiscus to be very careful in his wording of the letters:

> God Almighty will strengthen your arm so that you may conduct to a happy ending what you initiated out of zeal. As much as we can, all of us will help make a success of this affair. However, your Most Reverend Lordship must take care nevertheless in commencing the proceeding with the force of law not to introduce in your future letters anything contrary to formal and customary legal style, as it is called. For it often happens that even the tiniest clause may spoil an entire case, so that it is declared null and void if it comes before a higher judge.

When Reich received the three writs of disapproval, he was relieved to find an error in one of them, that of Scultetus, whose first name was given incorrectly. This gave him an excuse to send back the documents to Dantiscus for correction, saying that 'I am sending back all the letters because in one a serious scribal error must be corrected, and that cannot be done here. For the scribe wrote "Henry" instead of "Alexander" [Scultetus].'

When Dantiscus had the correction made and sent the letters back, Reich found yet another way out of presenting them to his colleagues. On 27 January 1539 he sent the letters back to Dantiscus, saying that there were not enough canons present to constitute a quorum. This was the end of his correspondence with Dantiscus, for on 1 March 1539 Reich died, having successfully avoided an embarrassing confrontation with his dear friend Copernicus, to whom he bequeathed all his books.

Dantiscus had also been in contact with another of the canons, Paul Plotkowski, who on 23 March 1539 wrote to him about the three women who were involved with his colleagues, all of them having been ordered to leave Frauenburg:

> As regards the Frauenburg wenches, Alexander's hid for a few days in his house. She promised that she would go away together with her son. Alexander returned from Lubawa with a joyous mien; what news he brought I known not. He remains in his *curia* with his *focaria*, who looks like a beer-waitress tainted with every evil. The woman of Dr Nicolaus did send her things to Danzig, but she herself stays on in Frauenburg.

Dantiscus knew that Tiedemann Giese, now Bishop of Kulm, was a close friend of Copernicus, who in early July 1539 was planning to visit him in his palace at Lubawa. Having learned of this through his informants, Dantiscus wrote to Giese, asking for his help in dealing with the scandalous behaviour of Copernicus, whom he said was being led astray by Alexander Scultetus:

> He [Copernicus] is renowned and recognized far and wide, not only with distinction but also with admiration in many fields of fine writing. In his old age, almost at the end of his allotted time, he is still said to let his mistress in frequently in secret assignations. Your Reverence would perform a great act of piety if you warned the fellow privately and in the friendliest terms to stop this disgraceful behavior, and no longer let himself be let astray by Alexander, whom he declares to be all by himself outstanding in all respects among all our brothers, the officials and canons.

Giese responded to Dantiscus by return mail, saying that he did as he was requested. Then two months later, after seeing Copernicus, he wrote again to say what had transpired:

> I have talked earnestly to Doctor Nicolaus about the subjects specified in your Reverence's warning. I put the situation, just as it is, before his eyes. He seemed to be distracted not a little. For although he has always obeyed your Reverence's wishes without delay, he is still falsely accused by malicious persons of secret assignations, etc. For he denies having

seen her, except that she spoke to him in passing as she was leaving for the fair in Krolewiec. I was absolutely convinced that he was not as emotionally involved as most people think.

There is no record of any further warnings given to Copernicus about Anna, and so it would seem that he satisfied Dantiscus that he had ended his relationship with her.

When Dantiscus told Giese that Copernicus was being led astray by Scultetus, he had in mind more than the affair with Anna. For in the spring of 1539 Scultetus was charged with having Lutheran beliefs, and Dantiscus was concerned that he had influenced Copernicus in this regard as well as in his morals.

In May 1540 King Sigismund summoned Scultetus to appear before him in Krakow, charging him with immoral behaviour and heresy. The summons was ignored by Scultetus, whereupon Sigismund banished him from Poland. Scultetus fled and took his mistress and their child with him to Rome, where he had influential friends. A hearing was held in Rome concerning the charge of heresy, but with the support of his friends Scultetus was acquitted. Meanwhile his *curia* in Frauenburg was searched, probably on the orders of Dantiscus, and two popular Protestant tracts were found among his papers, both of them with many marginal notes written by Scultetus. The evidence was passed to King Sigismund and then to Rome, where Scultetus was charged with heresy and imprisoned.

Copernicus was thus not only deprived of the company of Anna but also of that of his three best friends in Frauenburg, with the transfer of Tiedemann Giese to Kulm, the death of Felix Reich, and the flight of Alexander Scultetus. This, together with the charges of immorality and suspicions of heresy, would have made him feel besieged and isolated in his last years, with the revolutionary researches of a lifetime virtually unnoticed except for vague rumours that had made their way to the intellectual centres of Europe. But then a totally unexpected visitor arrived in Frauenburg who would change everything, including a world view that had endured since antiquity.

CHAPTER 9
THE FIRST DISCIPLE

Late in May 1539 Copernicus received an unexpected visit from a young German scholar, Georg Joachim van Lauchen, who called himself Rheticus (1514–74), and although only 25, was already Professor of Mathematics at the Protestant University of Wittenberg. He was accompanied by his *familus*, Heinrich Zell, who the year before had completed a map of Prussia based on the work of Alexander Scultetus and Copernicus.

Rheticus was deeply interested in the new cosmology of Copernicus, who received him hospitably and permitted him to study the manuscript that he had written to explain his theories. During the next ten weeks Rheticus worked with Copernicus in studying the manuscript, which he then summarized in a treatise entitled *Narratio prima* (First Account), intended as an introduction to the Copernican theory. Rheticus was to spend more than two years with working with Copernicus, a collaboration that would lead to the publication of *De revolutionibus*.

Rheticus later wrote of how he had first heard of Copernicus and decided to make the long journey to visit the great astronomer:

> Finally, hearing the great fame of Dr Nicholas Copernicus in the far north, even though the University of Wittenberg had appointed me professor in those disciplines, I knew I should have no rest until I myself learned something of his teaching. And indeed I regret neither the expense, nor the long journey, nor any of the other hardship. Rather, I feel, I have reaped a great reward. For by means of a certain youthful audacity I was able to spur this eminent man on to communicate to

the whole world his theories regarding that subject earlier than might have been. And all learned minds will join in my assessment of these theories as soon as the books we now have in press in Nuremberg are published.

Rheticus was born in the little alpine town of Feldkirch, just south of the southern tip of Lake Constance, in what is now Austria. His father, Georg Iserin, the town doctor, had brought his Italian wife, Thomasina de Porris, and their daughter Magdalena to Feldkirch from Lombardy just before the birth of their son Georg Joachim. When Rheticus was 14 his father was beheaded for sorcery, and his mother was forced to revert to her maiden name, de Porris, whose German form, 'von Lauchen,' was later adopted by her son Georg Joachim. When Georg went to university he followed the humanist custom of the time by taking as his surname the Latinized form of the name of his native region, and so he called himself Rheticus, from the ancient Roman province of Rhaetia, which included the region around Lake Constance.

After his father's death, Rheticus was sent to the cathedral school of the Frauenmünster Church in Zurich, where he remained for three years. There he studied mathematics and other subjects with Oswald Myconius (1488–1522), an old friend of Erasmus who had invited the Swiss reformer Ulrich Zwengli to Zurich. While at the school Rheticus formed a close friendship with his classmate Conrad Gessner (1516–65), who would become the most renowned zoologist and botanist of his time. Towards the end of his studies in Zurich, Rheticus also met Paracelcus, the famous physician, botanist, alchemist and astrologer.

Rheticus moved back to Feldkirch in the autumn of 1531, and during that winter he formed a lifelong friendship with the new town doctor, Achilles Pirmin Gasser (1505–77), who was also an astronomer, historian and philosopher. Gasser instructed Rheticus in astrology and the art of prognostication, and also introduced him to the pioneering work on magnetism written in the thirteenth century by Petrus Peregrinus. Gasser had studied at the University of Wittenberg in the years 1522–8 and knew Melanchthon, to whom he wrote a letter recommending Rheticus as a promising student.

In 1533 Rheticus matriculated at the University of Wittenberg, from which he took his MA on 27 April 1536. His curriculum there included astronomy, which was part of the mathematics discipline, at which he excelled, and also Greek, which he studied under Philipp Melanchthon (1487–1560), Rector of the university and Luther's right-hand man in the Protestant Reformation. He became a protégé of Melanchthon, who later wrote that Rheticus was 'born to study mathematics.'

One of his classmates was Erasmus Reinhold (1511–53), who also took his MA degree in the spring of 1536. Their mathematics teacher had been Johannes Volmer, who died in the previous winter. The vacancy was filled by Reinhold, who that same spring was appointed by Melanchthon as Professor of *mathematicum superiorum*, to lecture on astronomy and mathematics. Shortly afterwards Melanchthon created a second chair of mathematics for Rheticus, who was appointed Professor of *mathematicum inferiorum*, lecturing on arithmetic and geometry. Rheticus also lectured on astronomy and astrology, distributing his own notes on the subjects, containing the horoscopes of a number of important persons, including Luther and Melanchthon.

Melanchthon was very pleased by the lectures of Rheticus, particularly those in astrology, and in September he invited him to his home for a talk. He told Rheticus that he was giving him a leave of absence, to begin immediately, so that he could go to study with some of the leading scholars of Europe, beginning with Johannes Schöner of Nuremberg, who would in turn introduce him to some of his friends, including Peter Apian (Petrus Apianus) of Ingolstadt and Joachim Camerarius of Tübingen, Melanchthon's closest friend. Melanchthon wrote to Camerarius on 13 October, telling him that Rheticus had set off to see their friends:

> Greetings! This youth is our professor of mathematics. He has a nature suitable to the arts and not abhorrent to the humanities. As he is above all an astrologer, he has a strong command over that to which he is dedicated. Now he has gone forth to confer with Schöner and Apian on certain themes. Our mathematician wanted to greet you; he truly loves you greatly, not only due to your virtue and doctrine but also because of our friendship. I tell you this and beseech you, so that you will embrace him. He may not reach you for awhile, for which reason I write less.

Rheticus was accompanied to Nuremberg by a 19-year-old undergraduate named Nicholas Gugler, who had been asked by Melanchthon to assist him throughout his leave. When they arrived in Nuremberg they went to the home of Johannes Schöner, who put them up as his guests for the whole of their stay.

Johannes Schöner (1477–1547) was born in 1477 in the German town of Karlstadt, and at the age of 17 he enrolled in the University of Erfurt, but left without taking a degree to be ordained as a Roman Catholic priest in Bamberg. He converted to Lutheranism and left the priesthood to get married in 1527, by which time he had been brought to Nuremberg by Melanchthon to head the new Gymnasium there and teach mathematics, a position he held until the year before his death in 1547.

Schöner was renowned as an astronomer, astrologer, cartographer and instrument maker. He was also a writer, editor, printer and publisher, having set up his own printing press and publishing house in Bamberg, where he also made terrestrial and celestial globes for the books he wrote on geography and astronomy.

After Schöner moved to Nuremberg he set up his printing press and publishing house there. He also collaborated with another printer-publisher in Nuremberg, Johannes Petreius (c.1497–1550), a graduate of Wittenberg University who, among others, published works by Luther, Melanchthon and Henry VIII of England. Schöner and Petreius collaborated on a number of books, 13 of them by Regiomontanus, including his completions of unfinished works of Peurbach, which now appeared for the first time in print. One of these was Regiomontanus' *De Triangulis omnimodis* (On Triangles of Every Kind), published by Schöner in 1533. Schöner described it on the title page as a contribution 'to the perfecting of astronomical knowledge,' which is impossible 'without instruction in these things [i.e., triangles].' And in the preface he wrote that 'No one can bypass the science of triangles and reach a satisfying knowledge of the stars.' This was one of the valuable books that Rheticus would present to Copernicus when he came to see him in Frauenburg. As Rheticus wrote of these works in his *Narratio prima*, addressing Schöner:

> When I was with you last year and watched your work and those of other learned men in the improvement of the motions of Regiomontanus and his teacher Peurbach, I first began to understand what sort of task and how great a difficulty it was to recall this queen of mathematics, astronomy, to her palace, and to restore the boundaries of her kingdom.

In 1515 Schöner published a work on geography, in which his representation of the celestial globe shows the newly discovered land mass across the Atlantic labelled as 'America,' one of the first printed globes to do so. Two years later he published a work on astronomy including a representation of a celestial globe, the first ever to appear in print. Then in 1521 he published a work on an astronomical instrument called the planetary equatorium, including a representation of one that he had made himself. An equatorium is a type of spherical astrolabe which, in effect, functions as a specialized analogue computer to solve mathematical problems involving the motions of the planets, which in Schöner's model are represented as moveable disks. As James Evans describes it: 'The essential feature of an equatorium is that it takes account of the non uniformity of the planet's motion but nevertheless eliminates the need for trigonometry in making predictions.'

The appearance of a comet in August 1531 led Schöner to publish, later that year, a work on comets by Regiomontanus. Two years later Schöner published his major work on astrology, *Horoscopium generale, omni regioni accomodum*, defending a subject he had always supported.

Schöner had studied astronomy under Bernhard Walther, who, as noted, had been a student and collaborator of Regiomontanus in Nuremberg. As Regiomontanus writes of his reasons for moving to Nuremberg:

> I have chosen Nuremberg as my permanent home not only on account of the availability of instruments, particularly the astronomical instruments, on which the entire science of the heavens is based, but also on account of the very great ease of all sorts of communication with learned men living everywhere, since this place is regarded as virtually the center of Europe.

Regiomontanus remained in Nuremberg until 1475 when he returned to Rome to assist Pope Sixtus IV in his efforts to reform the calendar. He died there mysteriously on 6 July 1476, a month after his 40th birthday.

Walther helped Regiomontanus set up his own printing house as well as an observatory, the first in Germany. In 1472 Regiomontanus published Peurbach's *Theoricae novae planetarum*, the first printed textbook on astronomy, and he also made astronomical observations together with Walther., After the death of Regiomontanus the observations were continued by Walther, who in turn would later be assisted by Schöner. By the time of Walther's death in 1504, as Anton Pannekoek notes, 'he had made 746 measurements of solar altitude, and 615 determinations of the positions of the planets, moon and stars. It was first uninterrupted series of observations in the new rising of European science; a century later Tycho Brahe utilized them in his work.'

Three of Walther's observations were also used by Copernicus, who incorrectly attributed two of them to Schöner. These were all measurements of the positions of Mercury, which Copernicus records in *De revolutionibus*, V, 30, where he begins by complaining of the difficulties of observing this planet, particularly in the cold and misty environment of his own observatory, which was considerably farther north than those of the ancients:

> The ancients have directed us to the method of examining the motion of this planet, but they were favoured by a clearer atmosphere of a place, where the Nile – so some say – does not give out vapours as the Vistula does among us. For nature has denied that convenience to us who inhabit a colder region, where fair weather is rarer; and furthermore on account of the great obliquity of the sphere it is less frequently possible to see Mercury, as its rising does not fall within our vision at its greatest distance from the sun when it is Ares or Pisces, and its setting in Virgo and Libra is not visible; and it is not apparent in Cancer and Gemini at evening or early morning, and never at night, except when the sun has receded through the greater part of Leo. On this account the planet has made us take many detours and undergo much labour in order to examine its wanderings. On this account we have borrowed three positions those that have been carefully observed at Nuremberg.

He then describes the first of the observations, where the instrument that he refers to as an astrolabe is actually an armilla, or armillary sphere, which Walther had made using the one described by Ptolemy as a model:

> The first observation was taken by Bernard Walther, a pupil of Regiomontanus, in the year of Our Lord 1491 on the 9th of September, the fifth day before the Ides, 5 equal hours before midnight, by means of an astrolabe brought into relation with the Hyades. And he saw Mercury at $13\frac{1}{2}°$ of Virgo with a northern latitude of $1\frac{5}{6}°$; and at that time the

planet was at the beginning of its morning occultation, while during the preceding days its morning [elongation] had decreased continuously.

Copernicus goes on to describe the second and third observations, which he attributes to Schöner, who would have made them when he was working with Walther:

> The second was taken by John Schöner in the year of Our Lord 1504 on the 5th day before the Ides of January 6½ hours after midnight, when 10° of Scorpion was in the middle of the heavens over Nuremberg, when the planet was apparent at 3⅓°. Of Scorpio in the middle of the heavens over Nuremberg and the planet was apparent at 3⅓° of Capricorn with a northern latitude of 45'. Now by our calculation the mean position of the sun away from the spring equinox was at 27°7' of Capricorn and a morning Mercury was at 23°42° west of that. The third observation was taken by John Schöner in the same year 1504 the 15th day before the Kalends of April, at which time he found Mercury at 26¹/₁₀° of Aries with a northern latitude of approximately 3°, while 25° of Cancer was in the middle of the heavens over Nuremberg – as seen through an astrolabe brought in relation with the Hyades, at 12½ hours after midday, at which time the mean position of the sun away from the spring equinox was at 5°39' of Ares, an evening Mercury was 21°17' away from the sun.

As Owen Gingerich points out, 'Copernicus relied almost entirely on printed sources for his information, and there is only a single case where we know for sure that he used a manuscript source.' The single exception is the manuscript containing the three observations of Mercury described above.

Schöner had heard of Copernicus and his revolutionary theory, probably through reading a copy of the *Commentariolus*. He seems to have suggested that Rheticus might visit the great astronomer, and when his young visitor agreed to do so he gave him, among other gifts, a copy of the Mercury observations that he and Walther had made, knowing that this data was hard to come by. Schöner did not publish these observations until 1544, the year after Copernicus died.

Copernicus used Schöner's data to correct some of the parameters in his heliocentric theory. But he did not have time to correct his planetary tables to bring them into agreement with his revised data, as Owen Gingerich noted in checking the original manuscript of *De revolutionibus*, remarking that 'The Book was still filled with inconsistencies not yet ironed out.'

The values of the Mercury data used by Copernicus differed slightly from those published by Schöner in 1544. This work included the astronomical observations of Regiomontanus and Walther, as well as those that Schöner had made with Walther. Schöner also published manuscripts of Regiomontanus which had been in the possession of Walther, most notably *De triangulis omnimodis*, which he issued in 1533. These included 13 of the works that Regiomontanus had intended to print and publish, which he listed in a circular letter sent to his friends, but his untimely death prevented him from doing so. Anton Pannekoek gives a list of these works:

> The list of 22 items, all in Latin, mostly editions of ancient astronomers and mathematicians, includes Ptolemy's *Geography and Astronomy* ([...] in new translation); also Archimedes, Euclid, Theon, Proclus, Apollonius and others, followed by his own works, almanacs and minor writings. He began by publishing the planetary theory of his teacher Peurbach and the astronomical work by Manilius; after this carefully computed almanacs in Latin and German appeared. He won great fame with his *Ephemerides*, in which the position of the sun, moon and planets had been computed for 32 years, from 1475 to 1506.

Some of these works were included in the three bound books that Rheticus presented to Copernicus when he came to visit him in Frauenburg. The three books contained a total of six works, two of which were in the first bound volume, three in the second and one in the third. The works in the first volume were a 1533 edition of Euclid's *Elements*, and Schöner's edition of Regiomontanus' *De triangulis omnimodis*. The three works in the second volume were all published by Johnannes Petreius in 1534–5; they were Petrus Apianus' *Instrumentum primi mobilis*, Geber's *De astronomia*, and Witelo's *Perspectiva*. The third volume comprised a single work, the 1538 Greek edition of Ptolemy's *Almagest* (*Syntaxis*, in Greek).

Copernicus did not have any of these books, although he did have a Latin translation of Euclid, the *Elementa geometrica*, published in Venice in 1482, that he had acquired during his student days in Italy, and also the Latin translation of Ptolemy by Gerard of Cremona, published in 1515.

The First Disciple

Petrus Apianus (1495–1522) is renowned for his *Cosmographicus liber* (1524), a work on astronomy and navigation that would appear in at least 30 reprints in 14 languages and remain popular until the end of the sixteenth century. The terrestrial globe in this work was one of the first to label the New World as America. The fame of this work led to his appointment in 1527 at the University of Ingolstadt as mathematician and printer. That same year he published a handbook on commercial arithmetic in German called the *Rechnung*, one of a number of early algebraic works that appeared in Germany at that time. This work is notable for the fact that on the cover there is a diagram of the so-called 'Pascal triangle,' a triangular array of the coefficients of a binomial expansion, almost a century before Pascal was born.

Apianus observed a comet in 1531 and found that its tail always pointed away from the sun, the first to make this discovery, which is illustrated by a drawing in a book in German that he published in 1532. He published three manuals on sundials and other astronomical instruments, which he made and sold in the shop where he had his printing-press and publishing house. In his *Introduction to Geography*, published in 1533, he emphasized the importance of triangles to celestial as well as terrestrial measurements. His *Instrumentum primi mobilis*, published in 1534, is a work on trigonometry containing tables of the sine function. In 1540 he published a work called *Astronomicum Caesareum*, dedicated to the emperor Charles V, who appointed him Court Mathematician.

Geber is the Latin name of Jabir ibn Aflah, the twelfth century Andalusian astronomer. His major work, *Islah al-Majisti* (Correction of the Almagest), was translated into Latin by Gerard of Cremona as *De Astronomia*. This was the first criticism of Ptolemy's *Almagest* to appear in Western Europe and it was widely known. The original work in Arabic was one of the Islamic manuscripts available at the University of Krakow when Copernicus was a student there.

Jabir was critical of Ptolemy for his lack of rigour, and, in the case of the upper, that is outer, planets, for 'taking the center of the deferent to be halfway between the "equant" and the center of the universe without proof.' Jabir tried to improve the mathematical precision of the *Almagest* by using methods in spherical trigonometry by a group of tenth-century Arabic mathematicians, none of whom he credits. According to Geronimo Cardano, the sixteenth-century Italian mathematician, much of the material in Regiomontanus' *De triangulis*

(On Triangles) was taken directly and without credit from Jabir's work. It has been suggested that the spherical trigonometry that Copernicus outlines in the first part of *De revolutionibus* was also inspired by Jabir.

Witelo, the first European scientist of Polish origin, studied at the University of Padua (*c.*1260) and then moved to Viterbo. There he became a close friend of the great translator William of Moerbeke, to whom he dedicated his major work, *De perspectiva*, completed in *c.*1270–8. *De perspectiva* is based largely on the *Book of Optics* of the Arabic scientist Ibn al-Haytham, which Copernicus possessed in a Latin translation printed in Venice in 1485.

Schöner introduced Rheticus to his fellow-publisher Johannes Petreius and they became good friends. Petreius probably gave Rheticus the books that he had printed so that his young friend could then present them to Copernicus, whose *De revolutionibus* he would later publish.

While Rheticus was in Wittenberg he also met Andreas Osiander (1498–1552), a young priest who had converted to Lutheranism and became one of its most militant spokesmen. Osiander had converted the Grand Duke Albrecht of Prussia in 1525, although the conversion did not become official for a few years. He was very interested in the mathematical sciences, and when he learned that Rheticus was in town he sought him out, beginning a relationship that would end in bitterness and controversy.

Late in 1538 Rheticus left Nuremberg, and travelled to Ingolstadt, where he paid a visit to Petrus Apianus (Peter Apian), the well-known map-maker. Then early in 1539 he went on to Tübingen to visit the famous humanist Joachim Camerarius (1500–74), a close friend of Melanchthon.

Camerarius had been commissioned by Duke Ulrich of Württemberg in 1535 to reorganize the University of Tübingen, where he was Professor of Greek. Then in 1541 he went on to reorganize the University of Leipzig, where he spent most of the rest of his life. He translated into Latin many of the Greek classics, including works of Homer, Herodotus, Sophocles, Demosthenes, Lucian and Ptolemy. In 1535 he produced the first printed Greek edition of Ptolemy's astrological work, the *Tetrabiblos*, along with his annotated Latin translation of Books I and II as well as portions of Books III and IV. He published more than 150 books on a wide variety of subjects, including a work on

The First Disciple

Euclid in Latin, some of which were printed in Nuremberg by Petreius. Rheticus established a lifelong friendship with Camerarius as well as with his son Joachim Jr. (1534–98).

Rheticus returned to Wittenberg for a week early in May, when he was granted an extension of his leave of absence by the new academic head of the university, Casper Cruciger, who had recently replaced Melanchthon. At the same time he arranged to replace his *famulus* Nicolas Gugler, who had remained behind in Tübingen. His new secretary was Heinrich Zell (1518–64), a brilliant young man who had already embarked on his career as a cartographer.

Rheticus then set off for Frauenburg with Zell, sending a note from Posen on 14 May to Schöner telling him that he was on his way to see Copernicus. He refers to this letter at the beginning of the *Narratio prima*, which is addressed to 'The Illustrious John Schöner, as to his own revered father G. Joachim Rheticus sends his greetings.

> On May 14th I wrote you a letter from Posen, in which I informed you that I had undertaken a journey to Prussia, and I promised to declare, as soon as I could, whether the actuality answered to report and to my own expectation. However, I have been able to devote scarcely ten weeks to mastering the astronomical works of the learned man to whom I have repaired, for I had a slight illness and, on the honorable invitation of the most reverend Tiedemann Giese, bishop of Kulm, I went with my teacher [Copernicus] to Lubawa and there rested from my studies for several weeks. Nevertheless, to fulfill my promises at last and to gratify your desires, I shall set forth, as briefly and clearly as I can, the opinions of my teacher on the topics which I have studied.

Tiedemann Giese was Copernicus' closest friend, and they had worked together as canons of the Warmia chapter for more than 30 years. Giese was a distinguished scholar and had published a number of books, including a work on Aristotle. He carried on an active correspondence with Melanchthon and the great humanist Desiderius Erasmus of Rotterdam (1466–1536). His last exchange of letters with the latter took place shortly before Erasmus died, on 12 July 1536. As Erasmus wrote in his last letter to Giese:

> Your letter full of erudition and talent made me vividly regret being unable to meet your expectation. Your friend Eberhard will tell you in what a state I am, almost always lying in bed, with a health so much

altered. I have to turn down any studies without which life has no charm, even if I were fine. That is why, excellent man, if any excuse appears acceptable to you, at least forgive a dying man. Farewell, Basel, the 6th of June in the year 1536. Your Erasmus of Rotterdam, with his ill hand. To the distinguished master Tiedemann Giese, in Prussia.

Giese was also deeply interested in astronomy, as is evident from what Rheticus writes about him in the *Narratio prima*. This was in an appendix entitled 'In Praise of Prussia,' which was included in the first edition of the work, printed at Danzig in 1540 but omitted in some later editions. Here Rheticus writes of the kindness of two patrons he had met in Prussia, beginning with Giese:

> One of them is the illustrious prelate whom I mentioned at the outset, the Most Reverend Tiedemann Giese, bishop of Kulm. His Reverence mastered with complete devotion the set of virtues and doctrine, required of a bishop by Paul [...] In addition, the benevolent prelate deeply loves these studies [astronomy] and cultivates them earnestly. He owns a bronze armillary sphere for observing equinoxes, like the two somewhat larger ones which Ptolemy says were at Alexandria and which learned men from everywhere in Greece came to see. He has also arranged that a gnomon truly worthy of a prince should be brought to him from England. I have examined this instrument with the greatest pleasure, for it was made by an excellent workman who knew his mathematics.

Rheticus then goes on to write about the second of his two patrons in Prussia, 'the esteemed and energetic John of Werden, burgrave of Neuenburg, etc., mayor of the famous city of Danzig,' whom he says was praised by Copernicus as being comparable to 'Homer's Achilles, as it were.' Rheticus was grateful to John of Werden for the hospitality that the mayor would show him when he went to Danzig to arrange for the publication of the *Narratio prima*.

The visit to Lubawa began in mid-July 1539, just after Bishop Dantiscus had written to Tiedemann Giese urging him to warn Copernicus to sever his ties with Anna Schilling. Achacy von Trenck, who had been appointed Administrator of the Warmia chapter, visited Lubawa while Copernicus and Rheticus were there. He wrote to Dantiscus about his visit, in a letter dated 13 September 1539, reporting on the bishop's efforts to have Copernicus and his fellow canons Alexander Scultetus and Leonard Niederhoff break off relations with their mistresses:

> In the proceeding against Alexander in my opinion your Paternal Reverence will be able to settle hardly anything definite at the present time, since the housekeep has not been seen in Frauenburg since the day she left. I visited the Reverend Bishop of Kulm [Giese], as I had promised him a long time ago, and consulted with him how the Dean [of the Warmia Chapter, Leonard] Niederhoff could be persuaded [...] But it is useless talking to him because he always sings the same song. When Doctor Nicolaus, whom I found at Lubawa, heard his housekeeper mentioned, he declared that he would never receive her in his house nor do anything further in this case. I knew that he was advised by the Reverend Bishop of Kulm to do so, I hope not in vain, since his age and prudence can readily keep an upright, good man from actions of this kind in the future.

According to Rheticus, while they were in Lubawa Giese persuaded Copernicus to publish his astronomical work, something that the bishop apparently had been trying to do for some time.

> He realized that it would be of no small importance to the glory of Christ if there existed a proper calendar of events in the Church and a correct theory and explanation of the [celestial] motions. He did not cease urging my teacher, whose accomplishments and insights he had known for many years, to take up this problem, until he persuaded him to do so.

Rheticus writes of how Copernicus agreed to compose new astronomical tables, but he pointed out that this would require the use of hypotheses that ran counter not only to accepted theory but would seem to contradict common sense as well. Here he mentions John Angelus, a professor of astronomy at the universities of Ingolstadt and Vienna, who had compiled a set of astronomical tables that apparently remained unpublished at the time of his death in 1512.

> Since my teacher was social by nature and saw that the scientific world also stood in need of an improvement of the motions, he readily agreed to the entreaties of his friend, the reverend prelate. He promised that he would draw up new tables with new rules and that if his work had any

value he would not keep it from the world, as was done by John Angelus. But he had long been aware that in their own right the observations in a certain way required hypotheses which would overturn the ideas concerning the order of the motion of the motion and spheres that had hitherto been discussed and promulgated and that were commonly accepted and believed to be true; moreover the required hypotheses would contradict our senses.

Rheticus tells of how Copernicus, though agreeable to composing new planetary tables, was still reluctant to reveal the revolutionary theory that formed their basis:

> He therefore decided that should imitate the *Alfonsine Tables* rather than Ptolemy and compose tables with accurate rules but no proofs. In that way he would provoke no dispute among philosophers; common mathematics would have a correct calculus of the motions; but true scholars, upon Jupiter had looked with unusually favorable eyes, would easily arrive, from the numbers set forth, at the principles and sources from which everything was deduced. Just as heretofore learned men had to work out the true hypothesis of the starry sphere from the Alfonsine doctrine, so the entire system would be crystal clear to learned men. The ordinary astronomer, nevertheless, would not be deprived of the use of the tables, which he seeks and desires, apart from all theory. And the Pythagorean theory would be observed that philosophy may be pursued in such a way that its inner secrets are reserved for learned men, trained in mathematics, etc.

But Giese persisted, urging Copernicus to publish not only his planetary tables but the theory on which they were based, however revolutionary it might be. As Rheticus writes:

> Then His Reverence pointed out that such a work would be an incomplete gift to the world, unless my teacher set forth the reasons for his tables and also included, in imitation of Ptolemy, the system or theory and the foundations and proofs upon which he relied to investigate the mean motions [...] Moreover, contended the bishop, since the required principles and hypotheses are diametrically opposed to the hypotheses of the ancients, among scholars there would scarcely be anyone who would hereafter examine the principles of the tables and publish them after the tables had gained recognition as being in agreement with the truth. There was no place in science, he asserted, for the practice, he

asserted, frequently adopted in kingdoms, conferences, and public affairs, where for a time plans are kept secret until the subjects see the fruitful results and remove from doubt the hope that they will come to approve the plans.

Rheticus then concludes his account of Giese's successful efforts to persuade Copernicus to publish his revolutionary theory, paying tribute to the bishop for this gift to posterity:

By these and many other contentions, as I learned from friends familiar with the entire affair, the learned prelate won from my teacher a promise to permit scholars and posterity to pass judgment on his work. For this reason men of good will and students will be deeply grateful with me to His Reverence, the bishop of Kulm, for presenting this achievement to the world.

After he returned with Copernicus to Frauenburg, Rheticus immediately began to write the work that came to be called the *Narratio prima*, which he intended to be a clear and non-mathematical introductory summary of the Copernican theory. He does not mention Copernicus by name, referring to him throughout as 'my teacher.' Nevertheless, on the title page of the first edition of the *Narratio prima* the names Schöner and Copernicus are explicitly mentioned, while Rheticus does not give his own name. It reads, in translation:

TO THE MOST ILLUSTRIOUS GENTLEMAN MR. JOHANN SCHÖNER, CONCERNING THE BOOKS OF THE REVOLUTIONS Of the most learned Gentleman and Most distinguished Mathematician, The revered Doctor Mr. Nicolaus Copernicus of Torun, Canon of Warmia, By a certain youth Most zealous for Mathematics – A FIRST ACCOUNT

Rheticus concludes the *Narratio prima* with a quote from a lost play by Euripides, 'The opinions of older men are better,' after which he ends with the words 'From my library at Frauenburg, 23 September 1539.' It is not known where the library was, but it was probably in the *curia* of Copernicus, where Rheticus would seem to have lived during his entire stay in Frauenburg.

CHAPTER 10
THE FIRST ACCOUNT

When Rheticus finished writing the *Narratio prima*, he and Heinrich Zell took it to Danzig to have it printed and published. They brought the manuscript to the printing shop of Franciscus Rhodes, to whom Rheticus would have been introduced by John of Werden, and he agreed to publish it. Rheticus then returned to Frauenburg, probably leaving Zell behind in Danzig to supervise the printing of the manuscript and proofread the pages as they were printed.

The first leaves of the *Narratio prima* came off the press in mid-February 1540, and Rheticus sent copies to both Melanchthon and Osiander in Wittenberg. When the entire book was printed and ready for sale in Danzig, in April 1540, Rheticus sent copies to his friends in Nuremberg, including Osiander, Schöner and Petreius, as well as to Achilles Gasser in Feldkirche.

Rheticus had already written to Osiander to say that he was working with Copernicus. His letter has been lost, but a copy of the closing part of Osiander's reply, dated 13 March 1540, has survived. Osiander had by then read the first leaves of the *Narratio prima* that Rheticus sent him, and in his reply he discusses the speculations that Rheticus had made on the astrological significance of the Copernican theory (which Copernicus himself did not share). At the close of his reply Osiander asks Rheticus for an introduction to Copernicus so that he could correspond with him:

> But now this is enough about these topics. What remains is that I ask you over and over again, just as you offer me your friendship, in the

same way exert your efforts so as to obtain the friendship of this man [Copernicus] for me too. For the time being, I have not been able to write to him. I also did not even want to, being certain that you would not conceal these trifles of mine from him. I [heartily honor] him for his intellectual talents and his way of life. I also congratulate myself that up to the present time I have refrained from publishing my own material, and have not deprived him of the glory he deserves. Farewell. Think kindly of this confused and disorderly letter of a very busy man.

After Osiander received several copies of the full text of the *Narratio prima*, he wrote back to thank Rheticus on 20 April 1541. Only the opening lines of the letter have survived: 'Greetings, most excellent man and very dear friend. I have received several copies of your astronomical [*First*] *Report*. They have pleased me very much. The book has very clear introductory discussions of topics to be expected hereafter, such as.'

After reading the work thoroughly Osiander wrote to Copernicus. The letter is lost, as is the reply from Copernicus, dated 1 July 1540, which, for some reason, did not reach Osiander until March 1541. Only part of Osiander's reply to this letter, dated 20 April 1541, survives, but it indicates that Copernicus was afraid that his theory would be criticized by both Aristotelians and theologians:

> I have always felt about hypotheses that they were not articles of faith but the basis of computation. Thus, even if they are false, it does not matter, provided that they produce exactly the phenomena of the motions. [...] It would therefore appear to be desirable for you to touch upon this matter somewhere in an introduction. For in this way you can mollify the peripatetics and theologians, whose opposition you fear.

Osiander also wrote to Rheticus on 20 April 1541, sending the letter along with the same courier who was taking his letter to Copernicus. The only surviving part of this letter complements the letter he sent to Copernicus:

> The peripatetics and theologians will be readily placated if they hear that there can be different hypotheses for the same apparent motion; that the present hypotheses are brought forward, not because they are in reality true, but because they regulate the computations of the apparent and combined motion as conveniently as may be; that it is possible for someone else to devise different hypotheses; that one man

may conceive a suitable system, and another a more suitable, while both systems produce the same phenomena of motion; that each and every man is at liberty to devise more convenient hypotheses; and that if he succeeds he is to be congratulated. In this way they will be diverted from stern defense and attracted by the charm of inquiry; first their antagonism will disappear, then they will seek the truth in vain by their own devices, and go over to the opinion of the author.

After Johannes Petreius received a copy of the *Narratio prima* he wrote to Rheticus in August 1540, complementing him on the devotion to scholarship that had led him to bring the work of Copernicus to the world's attention:

> Now a year has passed since you were here with us, not that you could procure goods, like merchants for the sake of riches, but so that you could come to know the most distinguished man of our city, and the most justly meritorious in learning, Johann Schöner, and confer with him on the motions which the wonderful celestial bodies display. You considered this the most profitable commerce, and you thought yourself treated admirably because our Schöner, by virtue of his extraordinary kindness, was not only delighted by your talent, but also generously imparted what he believed would be beneficial to you in this system of learning. This desire for learning drew you to the farthest corner of Europe, to a distinguished gentleman [Copernicus], whose system, by which he observed the motions of the heavenly bodies, you related to us in a splendid description. Although he does not follow the common system by which these arts are taught in the schools, nevertheless I consider it a glorious treasure if some day through your urging his observations will be imparted to us, as we hope will come to pass.

Rheticus also gave copies of the *Narratio prima* to Tiedemann Giese, who sent one to Duke Albrecht of Prussia, along with a letter, dated 23 April 1540:

> Since the astronomical speculations of the worthy gentleman Nicolaus Copernicus, canon of Frauenburg, because of their extraordinary novelty, seem strange to everyone, and have also stirred up a highly learned mathematician from the University of Wittenberg to come to this land of Prussia that he might investigate their foundation and cause – and now, in advance of the appearance of the new astronomy of that gentleman doctor, has established in the form of a short book and preliminary

announcement, in which he has not neglected also to praise this land and make glorious mention of the name of Your Princely Eminence – I am sending along, lest Your Princely Eminence should not have received this little book, a copy of it as current information and with the earnest request that Your Princely Eminence might look graciously upon this highly learned guest on account of his great knowledge and skill, and grant him your gracious protection.

Gemma Frisius (1508–55), the famous Dutch physician, cartographer, mathematician and instrument maker, had read a copy of the *Narratio prima* soon after it appeared. On 20 July 1541 he wrote to his friend and patron Bishop Dantiscus, enthusing over the appearance of a new astronomer whom he hoped would rectify the deficiencies of the Ptolemaic model. As he wrote to Dantiscus: 'And so, if that author of yours [Copernicus] should achieve a restoration [of this science] to a state of soundness and good repair (which my mind has been greatly anticipating since I received the advance account that was sent), then would that not offer up a new earth, new heavens, and a new universe?'

However, Frisius was not particularly interested in the heliocentric theory of planetary motion, but, rather, the 'very precise calculations' that the *Narratio prima* led him to expect. As he wrote to Dantiscus: 'I do not argue about the hypotheses which that [astronomer] uses in his account, whatever they be, or however much truth they may possess. Nor does it concern me whether the earth is said to revolve, or whether it stand still.'

The *Narratio prima* was received with such favourable interest that a second edition was published at Basel in July 1541, under the supervision of Achilles Gasser, Rheticus' old tutor. Here, unlike the first edition, the name of Rheticus appeared on the title page. Gasser also added as a preface a letter he had written to his friend Georg Vögelin, a physician and philosopher in Constance, to whom he had sent a copy of the *Narratio prima*. Here, following an introductory sentence, Gasser describes the book to Vögelin:

> The book certainly departs from the manner of teaching practiced so far. As a whole it may run contrary to the usual theories of the schools and may even sound (as the monks would say) heretical. Nevertheless, what it seems to offer is the restoration – or rather, the rebirth, of a true system of astronomy. For in particular it makes highly evidential

claims concerning questions that have been sweated over and debated all across the world not only by very learned mathematicians but also by the greatest philosophers: the number of the heavenly spheres, the distance of the stars, the rule of the sun, the position and course of the planets, the exact measurement of the year, the specification of solsticial and equinoctial points, and finally the position and motion of the earth, along with other such difficult matters.

Gasser goes on to ask Vögelin to examine the work critically and pass it on 'to all who cherish mathematics [...] Recommend that they read it too. Not only will this hasten the appearance of a further, fuller account. But a greater stream of requests will reach its author [...] imploring him to allow the delivery of his whole work to us.'

The preface also included a ten-line poem that Vögelin had written in praise of the work, expressing his astonishment at the originality of the new theory, and suggesting that its novelty would arouse hostility among conservatives.

The preface of the *Narratio prima*, addressed to Schöner, begins with an account of Rheticus' journey to Prussia and his stay in Frauenburg up to his visit to see Tiedemann Giese in Kulm with Copernicus, whom he then describes, referring to him throughout as 'my teacher':

First of all I want you to be convinced, most learned Schöner, that this man whose work I am now treating is in every field of knowledge and in mastery of astronomy not inferior to Regiomontanus. I would rather compare him with Ptolemy, not because I consider Regiomontanus inferior to Ptolemy, but because my teacher shares with Ptolemy the good fortune of completing, with the aid of divine kindness, the reconstruction of astronomy which he began, while Regiomontanus – alas, cruel fate, departed this life before he had time to erect his columns.

He continues, saying that 'My teacher has written a work of six books in which, in imitation of Ptolemy, he has embraced the whole of astronomy, stating and proving individual propositions mathematically and by the geometrical method.' The first book, as Rheticus describes it, 'contains the general description of the universe and the foundations by which he undertakes to save the appearances and the observations of all ages,' to which he added all of the geometric and trigonometric principles that were necessary for his analysis.

He goes on to say that the second book contains 'the doctrine of the first motion,' meaning the apparent daily rotation of the stars. The third book, he says, deals with the motion of the sun and the length of the year, all of which 'depends, in part, on the motions of the stars'; the fourth with the motion of the moon and the occurrence of eclipses; the fifth with the motion of the planets; the sixth with latitudes.

Rheticus says that he has 'mastered the first three books, grasped the general idea of the fourth, and begun to conceive the hypotheses of the rest.' He tells Schöner that he thought it unnecessary to describe the first two books, partly because he intends to do this in a second account (never written), and 'partly because my teacher's doctrine of the first does not differ from the common and received opinion,' except for the construction of new tables. 'Therefore,' he says, 'I shall set forth clearly to you, God willing, the subjects treated in the third book together with the hypotheses of all the remaining books, so far as at present with my meager mental attainments I have been able to understand them.'

Rheticus begins his discussion of the third book with a section entitled 'The Motion of the Fixed Stars.' Here he gives an account of Copernicus' determination that the period of precession of the equinoxes was 25,816 years, a value he found by comparing the observations of earlier astronomers with his own, beginning with those he made when he assisted Dominico Maria de Novara in Bologna.

The next section is entitled 'General Consideration of the Tropical Year.' Here he explains that the measurement of the uniformity of the solar year can be more correctly measured with respect to the stars than to the equinoctial points, because of what Copernicus believed to be the non-uniformity of the precession of the equinoxes. As he writes: 'Realizing that equality of motion must be measured by the fixed stars, my teacher carefully investigated the sidereal year. He finds that it is 365 days, 15 minutes, and about 24 seconds and that it has always been of this length from the time of the earliest observations.'

Rheticus discusses 'The Change in the Obliquity of the Ecliptic' in the next section. The obliquity of the ecliptic can be determined by measuring the zenith distance of the sun at the summer and distance solstices, and then bisecting the arc between the two tropics. Copernicus believed that the obliquity was changing periodically, and, according to Rheticus, he concluded 'that the entire period of the change in the obliquity is 3,434 Egyptian years.'

The First Account

At the beginning of the following section, on 'The Eccentricity of the Sun and the Motion of the Solar Apogee,' Rheticus says that 'Since every difficulty in the motion of the sun is connected with the variable and unstable length of the year, I must speak of the change in the apogee and eccentricity, in order that all measures of the inequality of the year may be enumerated. However, by the assumption of theories suitable to the purpose, my teacher shows that these causes are regular and certain.'

Then, after going into the details of Copernicus' observations and theories, Rheticus concludes with praise for his teacher's accomplishment, which he says gave him great comfort:

> My teacher further established that the velocity with which the center of the eccentric revolves is the same as that with which each value of the changing obliquity recurs. This discovery is indeed worthy of the highest admiration, since it is achieved with such great and remarkable agreement. The eccentricity was greatest about 60 BC, when the declination of the sun was also at its maximum. The eccentricity has decreased, moreover, in accordance with this single law, similar to no other. This and other like sports of nature often bring me great solace in the fluctuating vicissitudes of my fortunes, and gently soothe my troubled mind.

This sent Rheticus off on an astrological digression, in the section entitled 'The Kingdoms of the World Change with the Motion of the Eccentric.'

> I shall add a prediction. We see that all kingdoms have had their beginnings when the center of the eccentric was at some special point on the small circle. Thus, when the eccentricity of the sun was at its maximum, the Roman government became a monarchy; as the eccentricity decreased, Rome too declined, as through aging, and then fell. When the eccentricity reached the boundary and quadrant of mean value, the Mohammedan faith was established, another great empire came into being and increased very rapidly, like the change in the eccentricity. A hundred years hence, when the eccentricity will be at its minimum, this empire too will complete its period. In our time it is at its pinnacle from which equally swiftly, God willing, it will fall with a mighty crash. We look forward to the coming of our Lord Jesus Christ, when the center of the eccentric reaches the other boundary of mean value, for it was in that position at the creation of the world.

Rheticus returns from his flight of fancy in the latter part of this section, where he discusses the studies made by Copernicus concerning the solar motion and its influence on the length of the tropical year, a subject that he comes back to in the next section. Towards the end of that section, entitled 'Special Consideration of the Length of the Tropical Year,' he praises Copernicus for his efforts in this area:

> It was a difficult task to recover by this analysis the motions of the fixed stars and of the sun, and through the computation of these motion to gain a correct understanding of the length of the tropical year. A boundless kingdom in astronomy has God granted my learned teacher. May he, as its ruler, deign to govern, guard, and increase it, to the restoration of astronomic truth. Amen.

The next section is entitled 'General Considerations Regarding the Motions of the Moon, Together with the New Lunar Hypotheses.' He begins by discussing the two so-called inequalities of the lunar motion; both were known to Ptolemy and no more were discovered until after the time of Copernicus. The first inequality of the moon is its oscillation about the mean position in its orbit, vanishing at its apogee and perigee, its periodicity equal to that of the lunar anomalistic period, the time for it to return to the same latitude. This inequality is due to the eccentricity of the moon's orbit. The second inequality is a variation of the first, due to fluctuations in the eccentricity of the lunar orbit, vanishing around the times of new and full moons and reaching its maximum at quadrature, when the sun, earth and moon form a right triangle, where the right angle is at the moon.

Rheticus explains how Copernicus dealt with these two inequalities, starting with the first: 'He supposes therefore that the lunar sphere encloses the earth together with its adjacent elements, and that the center of the deferent is the center of the earth, about which the deferent revolves uniformly, carrying the center of the lunar epicycle.'

He then goes on to explain how Copernicus handled the second inequality by introducing a secondary epicycle.

> The second inequality, which appears in the distance of the moon from the sun, he saves as follows. He assumes that the moon moves on an epicycle of a concentric; that is, to the first epicycle, which in general is

The First Account

in evidence at conjunction and opposition, he joins a second small epicycle which carries the moon; and he shows that the ratio of the diameter of the first epicycle to the diameter of the second as 1097:237 [...] This is the device or hypothesis by which my teacher removes all of the aforementioned incongruities, and which satisfies all the appearances, as he clearly shows, and as can be inferred also from his tables.

Rheticus points out that this theory had eliminated the Ptolemaic equant, which, he says, Copernicus also dispensed with in his planetary theory:

Furthermore, most learned Schöner, you see here that in the case of the moon we are liberated from an equant by the assumption of this theory, which, moreover, corresponds to experience and all the observations. My teacher dispenses with equants for the other planets as well, by assigning to each of the three superior planets only one epicycle and eccentric; each of these moves uniformly about its own center, while the planet revolves on the epicycle in equal periods with the eccentric. To Venus and Mercury, however, he assigns an eccentric on an eccentric.

Then, for the first time, Rheticus mentions the new and revolutionary planetary theory of Copernicus, involving heliocentrism and a moving earth:

The planets are each year observed as direct, stationary, retrograde, near to and remote from the earth, etc. These phenomena, besides being ascribed to the planets, can be explained, as my teacher shows, by a regular motion of the spherical earth; by having the sun occupy the center of the universe, while the earth revolves instead of on the eccentric, which it has pleased him to name the great circle. Indeed, there is something divine in the circumstance that a sure understanding of celestial phenomena must depend on the regular and uniform motions of the terrestrial globe alone.

Then in the next section, entitled 'The Principal Reasons Why We Must Abandon the Hypotheses of the Ancient Astronomers,' Rheticus describes a number of astronomical phenomena that can be explained by having the sun at the centre and the earth moving in orbit around it along with the other planets, beginning with the precession of the equinoxes and the variation of the obliquity of the ecliptic, as well as the changes in the eccentricity of the sun and of the planetary orbits:

> In the first place, the indisputable precession of the equinoxes, as you have heard, and the change in the obliquity of the ecliptic, persuaded my teacher to assume that the motion of the earth could produce most of the appearances in the heaven, or at any rate save them satisfactorily. Secondly, the diminution of the eccentricity of the sun is observed, for a similar reason and proportionally, in the eccentricities of the other planets.

Rheticus then shows how the various phenomena involved in planetary motion point towards a heliocentric theory and the motion of the earth, particularly in the case of Mars:

> Thirdly, the planets evidently have the centers of their deferents in the sun, as the center of the universe [...] Venus and Mercury do not recede further from the sun than fixed, ordained limits because their paths encircle the sun; hence these planets necessarily share the mean motion of the sun [...] Mars unquestionably shows a parallax sometimes greater than the sun's, and therefore it seems improbable that the earth should occupy the center of the universe. Although Saturn and Jupiter, as they appear to us in their morning and evening rising, readily yield the same conclusion, it is particularly and especially supported by the variability of Mars when it rises. For Mars, having a very dim light, does not deceive the eye as much as Venus and Jupiter, and the variation of its size is related to its distance from the earth. Whereas at its evening rising Mars seems to equal Jupiter in size, so that it is differentiated only by its fiery splendor, when it rises in the morning just before the sun and is then extinguished in the light of the sun, it can hardly be distinguished from stars of the second magnitude. Consequently, at its evening rising it approaches closest to the earth, while at its evening rising it is farthest; surely this cannot in any way occur on the theory of an epicycle. Clearly then, in order to restore the motion of Mars and the other planets, a different place must be assigned to the earth.

Aside from the appendix 'In Praise of Prussia,' the rest of the *Narratio prima* is devoted to what Rheticus, in the title of the next section, calls 'Transition to the Explanation of the New Hypotheses for the whole of Astronomy.' He sums up his discussion in this section by saying that 'I shall proceed to set forth the remaining hypotheses of my teacher in an

The First Account

open and orderly manner, in an endeavor to throw some light on my previous statements.'

The first section dealing with the hypotheses is entitled 'The Arrangement of the Universe.' Here he says that Copernicus first 'established by hypothesis that the sphere of the stars, which we commonly call the eighth sphere, was created by God to be the region which would enclose within its confines the entire realm of nature, and hence it was created fixed and immovable as the place of the universe.' He then says that motion is relative, so that sailors on a ship out of sight of land in calm weather are not aware of any motion even they may be sailing at a good speed. 'Hence this sphere was studded by God for our sake with a large number of twinkling stars, in order that in comparison with them, surely fixed in place, we might observe the position and motions of the other enclosed spheres and planets.'

Rheticus goes on to say that 'Then, in harmony with these arrangements, God stationed in the center of the stage His governor of nature, king of the entire universe, the sun "To whose rhythm the gods move, and the world/Receives its laws and keeps the pacts ordained."' The verse is from *Urania*, by Giovanni Gioviano Pontano, a work printed at Venice in 1501, which Copernicus had in his library in Frauenburg and would have purchased while he was a student in Italy.

Rheticus then describes the arrangement of the other nested celestial spheres, beginning with that of the outermost planet:

> The other spheres are arranged in the following manner. The first place below the firmament or sphere of the stars falls to the sphere of Saturn, which encloses the spheres of first Jupiter, then Mars; the spheres of first Mercury, then Venus surround the sun; and the centers of the sphere of the five planets are located in the neighborhood of the sun. Between the concave surface of Mar's sphere and the convex of Venus', where there is ample space, the globe of the earth together with its adjacent elements, surrounded by the moon's sphere, revolves in a great circle which encloses within itself, in addition to the sun, the spheres of Mercury and Venus, so that the earth moves among the planets as one of them.

Rheticus then goes on a long and lyrical digression on the wonders of the universe now opened up by Copernicus, saying 'For if we follow my teacher, there will be nothing beyond the concave surface of the starry sphere for us to investigate, except insofar as Holy Writ has vouchsafed us knowledge, in which the road will again be closed to placing anything beyond this concave surface.'

This leads him to a discussion of the 'remarkable symmetry and connection of the motions and spheres, as maintained by the assumption of the foregoing hypotheses.' He contrasts the often contradictory hypotheses made by earlier astronomers with those of Copernicus, in which the planetary spheres move independently of one another in a uniform and harmonious manner that he had earlier likened to dancers in a chorus:

> However, in the hypotheses of my teacher, which accept, as explained, the starry sphere as boundary, the sphere of each planet advances uniformly with the motion assigned to it by nature and completes its period without being forced into any inequality by the power of a higher sphere. In addition, the larger spheres rotate more slowly, and, as is proper, those that are nearer to the sun, which may be said to be the source of motion and light, revolve more swiftly. Hence Saturn, moving freely in the ecliptic, revolves in 30 years, Jupiter in 12, and Mars in 2. The center of the earth measures the length of the year by the fixed stars. Venus passes through the zodiac in 9 months, and Mercury, revolving about the sun in the smallest sphere, traverse the universe in 80 days. Thus there are only six moving spheres which revolve about the sun, the center of the universe. Their common measure is the great circle which carries the earth, just as the radius of the spherical earth is the common measure of the circles of the moon, the distance of the sun from the moon, etc.

The next section is entitled 'The Motions Appropriate to the Great Circle and its Related Bodies. The Three Motions of the Earth: Daily, Annual, and the Motion in Declination.' The discussion of the first two motions, the earth's diurnal rotation on its axis and its annual orbital motion around the sun, are simple and straightforward. But the third, that of the motion of the earth's axis relative to the ecliptic and the fixed stars, is complicated, all the more so since it involves the precession of the equinoxes, the spurious theory of trepidation, and Copernicus' own mistaken theory of why the earth's axis maintains a constant direction, as well as his erroneous notions concerning variations in the precession, the obliquity and the eccentricity.

The First Account

Rheticus continues the discussion of the third motion in the next section, entitled 'Librations.' He then begins the last part of the *Narratio prima* proper, which he calls THE SECOND PART OF THE HYPOTHESES, divided into three sections, all devoted to the motion of the five planets, in other words, those other than the earth.

In the first section, after a wordy preamble, Rheticus says that the clearest evidence for the heliocentric theory comes from its ability to explain both the real and apparent motions of the planets in an orderly and harmonious manner:

> With regard to the apparent motions of the sun and moon, it is perhaps possible to deny what is said about the motion of the earth, although I do not see how the explanation of precession is to be transferred to the sphere of the stars. But if anyone desires to look either to the principal end of astronomy and the order and harmony of the system of the spheres or to ease and elegance and a complete explanation of the causes of the phenomena, by the assumption of no other hypotheses will he demonstrate the apparent motions of the remaining planets more neatly and correctly. For all these phenomena appear to be linked most nobly together, as by a golden chain; and each of the planets, by its position and order and every inequality of its motion, bears witness that the earth moves and that we who dwell upon the globe of the earth, instead of accepting its changes of position, believe that the planets wander in all sorts of motions of their own.

Rheticus goes on to say that the Copernican theory was able to explain the real and apparent motion of the planets without using the Ptolemaic equant, thus remaining true to the 'axiom that all the motions of the heavenly bodies are circular or are composed of circular motions.'

The second section is entitled 'The Hypotheses for the Motion in Longitude of the Five Planets'; the third and final one is 'The Deviation of the Planets from the Ecliptic.' Rheticus begins by saying that the so-called inequalities of planetary motion, meaning their apparent retrograde motion, is due to the orbital motion of the earth, which Copernicus explained through a combination of epicycles and eccentric circles: 'Just as my teacher chose to employ an epicycle for the moon, so for the purpose of demonstrating conveniently the order of the planets and the measurement of their motion, he has selected, for the three superior planets, epicycles on an eccentric, but for Venus and Mercury eccentrics on an eccentric.'

He then describes in great detail how the apparent motion of the planets is explained by the orbital motion of the earth, first assuming that the planetary orbits are all in the plane of the ecliptic and then accounting for the fact that they are all in somewhat different planes, in both cases also taking into account other differences in their orbits such as eccentricity.

Rheticus concludes the *Narratio prima* proper with an encomium for his teacher, saying that he had not formulated his revolutionary theory 'in a lust of novelty, from the sound opinions of the opinions of the ancient philosophers, except for good reasons and when the facts themselves coerce him.'

> Such is his time of life, such his seriousness of character and distinction in learning, such in short his loftiness of spirit and greatness of mind, that no such thought can take hold of him [...] But may truth prevail, may excellence prevail, may the arts ever by honored, may every good worker bring to light useful things in his own art, and may he search in such a manner that he appears to have sought the truth. Never will my teacher avoid the judgment of honest and learned men, in which he plans of his own accord to submit.

The appendix 'In Praise of Prussia,' appears in the Danzig edition of 1540 and the Basel edition of 1541, but it was omitted in some later editions. Rheticus wrote it to ingratiate himself with influential figures like Duke Albrecht of Prussia, Bishop Tiedemann Giese, and John of Werden, mayor of Danzig, whose patronage would help him get his work published and further his career. It is written in an extravagantly over-the-top allegorical style, as, for example, where he associates Prussia with Apollo and Venus and compares it to Rhodes:

> You might say that the buildings and the fortifications [of Prussia] are palaces and shrines of Apollo; that the gardens, the fields and the entire region are the delight of Venus, so that it might be called, not undeservedly, Rhodes. What is more, Prussia is the daughter of Venus, if you examine either the fertility of the soil or the beauty and charm of the whole land.

The First Account

In a less florid style, Rheticus writes of a comment made by Tiedemann Giese to Copernicus, warning him that he would receive the same kind of criticism as Ptolemy had, both from scholars and the unlearned:

> But if it is to be the intention and decision of scholars everywhere to hold fast to their own principles passionately and insistently, His Reverence warned, my teacher should not anticipate a fate more fortunate that that of Ptolemy, the king of this science. Averroës, who was in other respects a philosopher of the first rank, concluded that epicycles and eccentrics could not possibly exist in the realm of nature and that Ptolemy did not know why the ancients had posited motions of rotation. His final judgment is: 'The Ptolemaic astronomy is nothing, so far as existence is concerned; but it is convenient for computing the non-existent.' As for the untutored, whom the Greeks call 'those who do not know theory, music, philosophy, and geometry,' their shoutings should be ignored, since men of good will do not undertake any labors for their sake.

The reaction to the *Narratio prima* was so positive that a second edition of the work was published in 1541 at Basel. Edward Rosen writes that 'It is altogether likely that the welcome to the *Narratio prima* was the clinching argument that finally persuaded Copernicus to put his manuscript into the hands of a printer.'

CHAPTER 11
PREPARING THE REVOLUTIONS

After Rheticus left the manuscript of the *Narratio prima* with the printer at Danzig he returned to Frauenburg and rejoined Copernicus, who had already set to work on making the final revisions on *De revolutionibus* in preparation for its publication. Copernicus may have been assisted in the revisions by Rheticus, who in any event would have been reading the revised work in progress.

Rheticus was called back to the University of Wittenberg for a brief term in the winter of 1540–1, when he taught a course in mathematical astronomy. At the conclusion of the course he rejoined Copernicus, continuing their work on the revision of *De revolutionibus*.

On 6 April 1541 Duke Albrecht of Prussia invited Copernicus to Königsberg, asking him to help the court physicians to treat his counsellor, Georg von Kunheim, who was seriously ill. Copernicus arrived in Königsberg on 13 April, accompanied by Rheticus and Zell, and they remained until 3 May, by which time Kunheim had recovered. On 5 May the Duke wrote to the Chapter at Frauenburg, praising the 'God-given skill [...] [of his] dearly beloved honorable and learned Nicolaus Copernicus, doctor of medicine, etc., [the chapter's colleague] and dear, friendly older brother.'

During their stay in Königsberg, Rheticus spent much time discussing the Copernican theory with Duke Albrecht, who had received a copy of the *Narratio prima* from Tiedemann Giese. They also seem to have talked about cartography, probably together with Zell. For at the end of August Rheticus sent the Duke two letters from Frauenburg, including a work in German that he had written on chorography, or local geography, in this case a map of Prussia. The chorography seems to have been for the most part composed by Zell, who based it at least partly on the map of Prussia drawn ten years earlier by Copernicus and Alexander Scultetus. The following year Zell published in Nuremberg a large map of Prussia and the surrounding territory, which seems to be the same as the one presented to Duke Albrecht. According to Swerdlow and Neugebauer, the Zell work, based on that of Copernicus and Scultetus, 'is the first detailed map of the region ever made, and is an excellent example of mid-sixteenth century chorography.'

After their return to Frauenburg Copernicus would have continued on his revisions of *De revolutionibus*. The work of revision continued up until at least that summer, for in June 1541 Rheticus wrote in a letter from Frauenburg that his teacher was in good health and very busy writing.

Copernicus had originally written his work in seven major sections or books. Book I dealt with cosmology, ending with the brief reference to Aristarchus and with the 'Letter from Lysis,' a brief work on Pythagorean philosophy, both of which were eliminated. The original Book II, which was on mathematics, was then joined directly onto Book I.

One of the two letters that Rheticus sent in late August to Albrecht was a request that the Duke write two letters of recommendation for him. One was to the Elector of Saxony and the other to the University of Wittenberg, both requesting permission for Rheticus be granted an extension of his leave of absence so that he could deliver his teacher's book to the printer. But both requests were denied, and so Rheticus was forced to leave Frauenburg.

Late in September 1541 Rheticus left Frauenberg to return to Wittenberg, ending a stay of more than two years in Prussia. Rheticus and Copernicus had formed a deep bond of friendship during their time together, a young man at the very beginning of his brilliant career working together with an old man close to the end of an obscure life spent in revolutionary researches that were just beginning to become known to the world. But now the two of them knew that though they were unlikely ever see one another again, though Copernicus was confident that his lifelong work would be brought to fruition by Rheticus.

Sixteen years later, in a letter to the Emperor Ferdinand I, his patron, Rheticus reflected on this historic visit, which was to change the course of intellectual history:

> Nicolaus Copernicus, the Hipparchus of our age whom no one has words enough to praise adequately, was the first to understand the irregularity in the sphere of the fixed stars [...] In fact, after I had spent about three years in Prussia, the great old man charged me to carry on and finish what he, prevented by age and impending death, was unable to complete.

After Rheticus returned from Frauenburg to the University of Wittenberg in October 1541 he was promoted to professor of higher mathematics and elected dean of the faculty of arts. When the autumn term began he gave a course of lectures in astronomy, which he advertised as being an 'interpretation of Ptolemy.' He continued to serve as dean until the end of the winter term, in April 1542, when in his commencement oration he concluded by saying 'Let us rouse ourselves [...] to preserving the beneficial teaching for posterity.'

Meanwhile his friend and fellow professor of higher mathematics Emmanuel Reinhold had published a new, annotated edition of Peurbach's *Theoricae novae planetarum*. In his preface Reinhold mentioned that he knew of 'a modern astronomer who is exceptionally skillful, who has raised a lively expectancy in everybody; one hopes that he will restore astronomy [...] I hope that this astronomer, whose genius all posterity will rightly admire, will at long last come to us from Prussia.' Reinhold would have learned about the Copernican theory from Rheticus, and in the second edition of his book he mentioned Copernicus by name.

When Rheticus left Frauenburg he may have taken with him the completed version of *De revolutionibus*, which in any event Copernicus would have sent him during the course of the following year. What Rheticus certainly did have with him was a copy of Chapters 13 and 14 of Book I of *De revolutionibus*, which presented the theorems on plane and spherical trigonometry that Copernicus used in his analysis of celestial motions. He had this printed by Johannes Lufft in Wittenberg before June 1542 as *De lateribus et angulis triangulorum* (Sides and Angles of Triangles). He dedicated the book to the astronomer and instrument-maker Georg Hartmann (1489–1564) of Nuremberg, who had known

Andreas Copernicus in Rome, and who, along with Schöner, had urged Rheticus to visit Copernicus. Rheticus states in the dedication, addressing Hartmann, that 'The science of triangles, as you know, is useful in all things pertaining to geometry, but especially in astronomy.' He goes on to say that 'the most illustrious and highly learned Mr. Nicolaus Copernicus was an interpreter of Ptolemy' who 'wrote most effectively concerning triangles.' At the conclusion of the dedication Rheticus tells Hartmann of his admiration for Copernicus:

> Such a learned man as you will love this man equally for his brilliance and his learning, above all as regards the science of the heavens. In this he is comparable to the greatest creative minds of old. There has been no greater human happiness than my relationship with so excellent a man and scholar as [Copernicus] is. And should my own work ever make any contribution to the general good (to the service of which all our efforts are directed), it shall be owing to him.

According to Swerdlow and Neugebauer: this was 'Copernicus's first publication since the letters of Theophylactus thirty years before. Rheticus added a seven-place sine table at one minute intervals that appears to be a slight modification of Regiomontanus's table, which he could have obtained in Nuremberg when visiting Schöner, who himself had Regiomontanus's table printed by Petreius in 1541.'

During the following years Rheticus combined the ideas of Regiomontanus and Copernicus, together with his own methods, and wrote the most elaborate treatise on trigonometry composed up to that time. This was a massive two-volume work entitled *Opus palatium de triangulis* (Canon on The Science of Triangles), published in Leipzig in 1551. Carl B. Boyer describes this pioneering work: 'Here trigonometry really came of age. The author discarded the traditional consideration of the functions with respect to the arc of a circle and focused instead on the lines of a right triangle. Moreover, all six trigonometric functions now came into full use, for Rheticus calculated elaborate tables of all of them.'

The trigonometric tables were still unfinished when Rheticus died in 1574, but they were completed by his disciple Valentin Otho (c.1550–1605), who edited and expanded the treatise, which he published in 1595. This work was revised in turn by Bartholomaeus Pitiscus

(1561–1613), who had already published his own book on the subject, called *Trigonometria*, first printed at Heidelberg in 1595, translations of which appeared in England and France in 1614 and 1619, respectively, introducing the word 'trigonometry' to the English and French languages. Pitiscus is sometimes credited with inventing the decimal point, which appears in his trigonometrical tables and was subsequently used by John Napier in his pioneering papers on logarithms in 1614 and 1619.

Pitiscus, in his revision, made extensive use of papers that Rheticus had left behind, and published them in his *Thesaurus mathematicus*, printed at Frankfurt in 1613, whose full title in English is *Mathematical Treasury: or Canon of Sines for a Radius of 1,000,000,000,000,000 Units [...] as Formerly Computed by Georg Joachim Rheticus*. As Dennis Danielson remarks in his book on Rheticus, these 'computations produced tables so useful that they were only superseded by works appearing early in the twentieth century.'

Rheticus acknowledged his debt to Copernicus in an appendix to the tables of the Canon on *The Science of Triangles*. This is a dialogue between two characters, Philomathes and Hospes.

> Hospes: But this Rheticus – What sort of man is he? His name I have heard before, and now I see it written on the front of this little book.
> Philomathes: He indeed is the one who is now bringing us this fruit from the delightful gardens of Copernicus. For after his recent return from Italy he resolved to impart freely to students of mathematics everything he learned from that excellent old man, as well as everything he has acquired by his own effort, perseverance, and devotion.

Rheticus also acknowledged his indebtedness and devotion to Copernicus in a work he published in 1550. This was the *Ephemerides novae*, whose subtitle, in English, is *A Setting Forth of the Daily Position of the Stars [...] by Georg Joachim Rheticus according to the theory [...] of his teacher Nicolaus Copernicus of Torun*.

The *Ephemerides* had two prefaces, the first of which was addressed to Georg von Komerstadt, a nobleman in the service of the Prince-Elector Moritz of Saxony. There he tells of his pilgrimage to see Copernicus and his efforts to publish *De revolutionibus*: 'I searched for someone who could interpret the heavens and instruct me concerning the stars [...]

In Prussia I learned and grasped the splendid art of astronomy while staying with that distinguished man Copernicus. And to work out all these things, to unfold them and to embellish them, is more than life or the efforts of any one person to accomplish.'

Rheticus goes on to say that his work is part of a larger effort to establish the Copernican theory in the face of both the ignorant and the knowledgeable, as well as the conservative:

> There are those who, knowing nothing, will turn away in stupidity; others through their exceeding knowledge will, rashly carp at some bit they have seized upon; others will ridicule our work on account of its novelty. However, there is nothing one can write by way of defense against all these. Only a bulwark of support will suffice. What will convince them is not disputation but refutation.

In the second preface, 'The Author to the Reader,' Rheticus praises his teacher, fondly recalling their time working together:

> I remember myself being driven by juvenile curiosity. I wished to hasten to the stars' sanctuary. So agreeing with the very good and great Man, Copernicus, I sometimes blamed the painstaking attention to details. But he was bewildered by my soul's honest thirst, and with a soft arm, he used to exhort me to take my hand off the Tables. 'Personally,' he said, 'if I could get my truth from a sixth which is an increment of ten minutes, my spirit would exult as much as when was received the discovery of the formula of the ratios of Pythagoras.'

Rheticus goes on to talk of how Copernicus never let the detailed work of precise observations and analysis obscure his overall view of his researches and their aim:

> He wanted his researches to be above the average. That is why he avoided grindings not by inertia nor by fear of boredom. Some people seek and even require these little grains, like Peurbach in the subtlety of his table of eclipses. They can see in them all the care taken to locate the stars with precision. While they are impressed by the seconds, thirds, fourths, fifths, and little divisions, they forget the integer parts, not giving them a single look. And in the small interval of times of the 'Phenomena,' they are often wrong by hours, and, sometimes, entire days. There is an amazing fable of Aesop where an order is given to search for a lost cow.

It is found, but the men who are to bring it back, see little birds and go after them, forgetting the cow.

Rheticus had already, in the *Narratio prima*, given a very interesting and detailed description of how Copernicus combined the observations and theories of Ptolemy and others with his own to formulate his new planetary model:

> But when I see that my teacher always has before his eyes the observations of all ages together with his own, assembled in order as in catalogues; then when some conclusions must be drawn or contributions made to the science and its principles, he proceeds from the earliest observations to his own, seeking the mutual relationship which harmonizes them all; the results thus obtained by correct inference under the guidance of Urania he then compares with the hypotheses of Ptolemy and the ancients; and having made a most careful examination of these hypotheses, he finds that astronomical proof requires their rejection; he assumes new hypotheses, not indeed without divine inspiration and the favor of the gods; by applying mathematics, he geometrically establishes the conclusions which can be drawn from them by correct inference; he then harmonizes the ancient observations and his own with the hypotheses which he has adopted; and after performing all these operations he finally writes down the laws of astronomy…

Edward Rosen describes the atmosphere in Wittenberg at the time Rheticus published the *Sides and Angles of Triangles*:

> In Wittenberg, the mightiest fortress of anti-Catholic Lutheranism, Rheticus identified the author of *Sides and Angles of Triangles* geographically with his birthplace Thornn, not confessionally with the Roman Catholic Cathedral Chapter of Frauenburg. In the preface Rheticus explained that Copernicus 'wrote most learnedly about triangles while he was hard at work elucidating Ptolemy and expounding the theory of the [heavenly] motions.' 'Elucidating' (*illustrando*) Ptolemy, not contradicting him was what Rheticus depicted Copernicus as doing when the *Sides and Angles of Triangles* was published in Wittenberg.

Before Rheticus left Wittenberg for Frauenburg he was invited to dinner by Martin Luther. At that time Rheticus seems to have spoken about the new astronomical theory of the canon from Thorn. It was customary for some of the guests to write down what Luther said during these meals. The earliest edition of his *Tischreden*, or *Table Talk*, appeared in 1566, twenty years after his death. According to the *Table Talk*, at the dinner on 4 June 1539, shortly after Rheticus began his journey to Frauenburg, the conversation turned to the revolutionary theory of the obscure canon from Thorn, whereupon Luther called Copernicus a 'fool.' But in a later version of *Table Talk* the word 'fool' was removed and milder language was used.

The conversation concerned the idea of the relativity of motion, as for example a passenger on a moving coastal ship, who imagines himself to be still, while the shore seems to be moving in the opposite direction. Copernicus explains this idea in *De revolutionibus*, I, 5:

> For every apparent change in place occurs on account of the movement of the thing seen or of the spectator, or on account of the necessarily unequal movement of both. For no movement is perceived relatively to things moved in the same direction – I mean relative to the thing seen and the spectator. Now it is from the Earth that the celestial circuit is beheld and presented to our sight. Therefore, if some movement should belong to the Earth it will appear, in the parts of the universe which are outside, as the same movement but in the opposite direction, as though the things outside were passing over. For the daily revolution is especial in such a movement. For the daily revolution appears to carry the whole universe along, with the exception of the Earth and the things around it. And if you admit that the heavens possess none of this movement but that the Earth turns from east to west, you will find – if you make a serious examination – that as regards the apparent rising and setting of the sun, moon, and stars the case is so.

According to *Table Talk*, the guests were discussing the relativity of motion in connection with the Copernican idea of the moving earth, when Luther broke in to ridicule the idea as being contrary to Scripture:

> There was mention of a certain new astrologer who wanted to prove that the earth moves and not the sky, the sun and the moon. This would be as if somebody was riding in a cart or in a ship and imagined that he was standing still while the earth and the trees were moving. [Luther remarked,] 'So it goes now. Whoever wants to be clever must agree with nothing others esteem. He must do something of his own. This is what the fellow does who wishes to turn the whole of astronomy upside down. For even in these things that are thrown into disorder I believe the Holy

Scriptures, for Joshua commanded the sun to stand still and not the earth.'

Melanchthon also rejected the idea of a moving earth, probably following a conversation with Rheticus after he returned to Wittenberg. In a letter to a friend, on 16 October 1541, he wrote: 'Certain people think it is a wonderful achievement to elaborate such an absurd thing, like that Polish astronomer who moves the earth and holds the sun stationary. Certainly prudent government ought to restrain the smart alecks' insolence.'

Melanchthon's attitude made it clear to Rheticus that he would not be allowed to publish *De revolutionibus* in Wittenberg. He had been allowed to publish in Wittenberg Copernicus' *Sides and Angles of Triangles*, because it was a purely technical book and contained no theories or speculations, so that it did not run counter to traditional beliefs. And so Rheticus decided to publish *De revolutionibus* in Nuremberg with Johannes Petreius, who had already indicated that he was anxious to print it.

At the end of the winter semester, 1542, Rheticus applied for another leave of absence from the University of Wittenberg. As soon as his leave was approved he set off for Nuremberg, where in May of 1542 Rheticus delivered to Petreius a fair copy of the revised manuscript of *De revolutionibus*. By the end of the month the first two signatures, or folded leaves, were printed and Rheticus began proofreading them in June. By that time Copernicus had completed the preface and dedication of *De revolutionibus*, which he sent to Nuremberg, along with the letter that he had received six years before from Cardinal Schönberg, which he wanted to include as a headnote. As Edward Rosen notes, referring to the fact that Achilles Gasser had received a copy of the book when it was first printed: 'On folio 2 verso of the first signature Gasser noted that Copernicus' Dedication was composed at Frombork [Frauenburg] in Prussia in the month of June 1542. Copernicus' Dedication was addressed to the reigning pope, Paul III, whom he asked to protect him from calumnious attacks.'

Rheticus had already begun manoeuvring to obtain a professorship at the University of Leipzig. Joachim Camerarius had recently moved from Tübingen to become rector at the University of Leipzig and was trying to persuade Rheticus to join him there. Philipp Melanchthon wrote a letter to Camerarius supporting Rheticus' application for a professorship, commending him most warmly. But in a later letter he warned Camarerius that very definite terms should be specified in their young friend's contract, given the successive leaves of absence that Rheticus had taken while he was at the University of Wittenberg. Rheticus was appointed Professor of Mathematics at the University of Leipzig, where he was to begin teaching in mid-October 1542. An entry in the archives for 1542 notes that the university 'lured away from Wittenberg the eminent and most learned gentleman professor Joachim Rheticus.'

While *De revolutionibus* was at the printer, Rheticus left Nuremberg in the first half of June to visit his family and friends in his native region of Rhaetia, particularly Achilles Gasser in Feldkirche.

Rheticus returned to Leipzig by early July and resumed his proof-reading of *De revolutionibus* as the printed leaves came off the press. Then in early October he left to take up his professorship in Leipzig. Responsibility for proofreading and editing *De revolutionibus* was left to Andreas Osiander, who, as we have seen, had been in correspondence with both Rheticus and Copernicus about the heliocentric theory. It is not clear whether Osiander took on this task at the request of Rheticus or Petreius, or if he did it on his own initiative.

Osiander supervised the printing of the remainder of the main text of *De revolutionibus*, Books V and VI. He then presented to Petreius the front matter, which Copernicus had intended to comprise only his preface and dedication to Pope Paul III, along with the letter that he had received from Cardinal Nicolaus Schönberg, which was to be put before the preface as a headnote. Copernicus presumably added the letter from the Cardinal and his prefatory dedication to the Pope so as to give the impression that his work was not objectionable to the highest levels of the Catholic Church. But Osiander seems to have thought that this was not enough, and without consulting either Rheticus or Copernicus he interpolated at the beginning of the front papers an anonymous section entitled *Ad lectorem* (Address to the Reader). There he stated that the revolutionary concepts of heliocentrism and the earth's motion presented in *De revolutionibus* were only intended as a mathematical model and not a real description of nature.

To the Reader Concerning the Hypotheses of this Work:

> There have been widespread reports about the novel hypotheses of this work, which declares that the earth moves whereas the sun is at rest in

the center of the universe. Hence certain scholars, I have no doubt, are deeply offended and believe that the liberal arts, which were established long ago on a sound basis, should not be thrown into confusion. But if these men are will [ing] to examine the matter closely, they will find that the author of this work has done nothing blameworthy. For it is the duty of the scholar to compose the history of the celestial motion through careful and expert study. Then he must conceive and devise the causes of these motions or hypotheses about them. Since he cannot in any way attain to the true causes, he will adopt whatever suppositions enable the motions to be computed correctly from the principles of geometry for the future as well as the past. The present author has performed both these duties excellently. For these hypotheses need not be true or even probable. On the contrary, if they provide a calculus consistent with the observations, that alone is enough....

The *Ad lectorem* then goes on to say that since nothing is certain except when it is divinely inspired, then the hypotheses presented in *De revolutionibus* make no claim on absolute truth but should be considered as possibilities, supported by careful observations:

For this art [astronomy], it is quite clear, is completely and absolutely ignorant of the causes of the apparent non uniform motions. And if any causes are devised by the imagination, as indeed very many are, they are not put forward to convince that they are true, but merely to provide a reliable basis for computation. However, since different hypotheses are sometimes offered for one and the same motion (for example eccentricity and an epicycle for the sun's motion), the astronomer will take as his first choice that hypothesis which is easiest to grasp. The philosopher will perhaps seek the semblance of the truth, but neither of them will understand or state anything certain, unless it has been divinely revealed to him. Therefore alongside the ancient hypotheses, which are no more probable, let us permit these new hypotheses also to become known, especially since they are admirable as well as simple and bring with them a huge treasure of very skillful observations. So far as hypotheses are concerned, let no one expect anything certain from astronomy, which cannot furnish it, lest he accept as the truth ideas conceived for another purpose, and depart from the study a greater fool than when he entered it. Farewell.

After Petreius received the *Ad lectorem* from Osiander, he began printing *De revolutionibus*, which was finally published in mid-April 1543. The full title of the book is *De revolutionibus orbium coelestium libri VI* (Six Books

Concerning the Revolutions of the Heavenly Spheres). The publisher's blurb, probably also written by Osiander, was printed directly below the title. 'You have in this recent work, studious reader, the motion of both the fixed stars and the planets recovered from ancient as well as recent observations and outfitted with wonderful new and admirable hypotheses. You also have most expeditious tables from which you can easily compute the positions of the planets for any time. Therefore buy, read, profit.'

Petreius immediately sent the printed work to Rheticus. When Rheticus read the *Ad lectorem* he was shocked and infuriated. He deeply regretted not having stayed on at Nuremberg to supervise the final stages of printing the book, and he engaged in a bitter dispute with Johannes Petreius about this. Rheticus took a red crayon and drew a big X across the page containing most of the *Ad lectorem*, and he did the same in all of the copies that he sent out to his friends, as can be seen in surviving volumes of *De revolutionibus*.

Rheticus suspected that Osiander had written the *Ad lectorem*, and that he had interpolated it into the fair copy of *De revolutionibus* before giving it to Petreius for printing. Petreius vehemently denied that he had any part in the subterfuge, saying that he assumed that the *Ad lectorem* was part of the front papers of Copernicus' book.

Years later Osiander admitted that he was the author of the *Ad lectorem*, and that he had added it to the manuscript that he gave to Petreius without consulting anyone. He made this admission to the mathematician Peter Apian, who passed it on to his son Philip Apian (1531–89). Michael Maestlin (1550–1631), the teacher of Johannes Kepler, found a note regarding this admission among Philip Apian's books, whereupon he transcribed it into his own copy of *De revolutionibus*:

> Concerning this letter [*Ad lectorem*], I [Michael Maestlin] found the following among Philip Apian's books (which I bought from his widow) [...] 'On Georg Joachim Rheticus, the Leipzig professor and disciple of Copernicus, became embroiled in a very bitter dispute with the printer. The latter asserted that it had been submitted to him with the rest of the treatise. Rheticus, however, suspected that Osiander had prefaced it to the work, and declared that, if he knew this for certain, he would sort the fellow out in such a way that he would mind his own business and never again dare to slander astronomers. Nevertheless, [Peter] Apian told me that Osiander had openly admitted to him that he had added this [letter] as his own idea.'

Rheticus sent two copies of *De revolutionibus* to Tiedemann Giese and one to George Donner, an elderly canon of the Warmia chapter and close friend of Copernicus, who by that time was gravely ill. The earliest extant report of what proved to be Copernicus' last illness is in a letter, dated 8 December 1542, from Tiedemann Giese to George Donner:

> I was shocked about what you wrote about the impaired health of our venerable old man, our Copernicus. Just as he loved privacy while his constitution was sound, so, I think, now that he is sick, there are few friends who are affected by his condition. Yet we are all indebted to him for his uprightness and outstanding learning. I know that he has always reckoned you among those most faithful [to him]. I therefore ask you, if his condition so warrants, please to watch over him and take care of the man whom you cherish at all times together with me. Let him not be deprived of brotherly help in this emergency. Let us not be considered ungrateful towards this deserving man. Farewell.

On 29 January 1543, Bishop Dantiscus wrote to Gemma Frisius that Copernicus was paralyzed on one side of the body, apparently having suffered a stroke. A young nephew of Copernicus', Johannes Leutze, had already applied to be coadjutor, or successor, to his uncle's canonry in Warmia, and so it would seem that he was dying. Leutze's canonry was approved in May of that year, when Copernicus passed away. The exact date of his death is established by Tiedemann Giese in a letter he sent on 26 July 1543 to Rheticus, who had informed him of his shock and anger on having read the *Ad lectorem* that had been inserted at the beginning of *De revolutionibus*. Giese was equally indignant, and in the first half of his long letter he asked Rheticus to correct this outrageous distortion in any reprinting of the first leaves of the book.

> On my return from the royal wedding in Cracow [of Prince Sigismund Augustus of Poland with Elizabeth, archduchess of Austria] at Lubawa I found the two copies, which you had sent, of the recently printed treatise of our Copernicus. I had not heard about his death before I reached Prussia. I could have balanced out my grief at the loss of that very great man, our brother, by reading his book, which seemed to bring him back to life for me. However, at the very threshold I perceived the bad faith and, as you correctly label it, the wickedness of Petreius, which produced in me an indignation more intense than my previous sorrow. For who will not be anguished by so disgraceful an act, committed

> under the cover of good Faith [...] I have written to the City Council of Nuremberg, indicating what I thought had to be done in order to restore faith in the author. I am sending you the letter together with a copy of it, to enable you to decide how the affair should be managed on the basis of what has been started [...] If the first sheets are going to be printed again, it seems that you should add a brief introduction which would cleanse the stain of chicanery also from those copies which have already been dedicated.

Giese goes on to describe the last hours of Copernicus, after which he discusses changes and additions that Rheticus should make in a new edition of *De revolutionibus*, which he thought should be sent to the pope.

> I should like in the front matter also the biography of the author, tastefully written by you, which I once read. I believe that your narrative lacks nothing but his death. This was caused by a hemorrhage and subsequent paralysis of the right side on 24 May, his memory and mental alertness having been lost many days before. He saw his treatise only at his last breath on his dying day [...] I should like also the addition of your little tract, in which you entirely correctly defended the earth's motion from being in conflict with the Holy Scriptures. In this way you will fill the volume out to a proper size and you will repair the injury that your teacher failed to mention you in his Preface to the treatise [...] As for the copies of the treatise which you sent to me, I am deeply grateful to the donor. These copies will serve me as a permanent reminder to preserve the memory not only of the author, whom I always cherished, but also of you. Just as you proved to be an energetic assistant to him in his labors, so now you have helped us with your effort and care lest we be deprived of the enjoyment of the finished work. It is no secret how we all owe you for this zeal. Please let me know whether the book has been sent to the pope; for if this has not been done, I would like to carry out this obligation to the deceased. Farewell.

The biography of Copernicus to which Giese refers, which is mentioned by no other source, is lost. The tract on the compatibility of the earth's motion and Holy Scriptures that he mentions has survived, though it was printed anonymously and not discovered and identified as Rheticus' work until the early 1980s.

Copernicus was buried beneath the pavement in the nave of Frauenburg Cathedral, near the tomb of his uncle Lucas Watzenrode.

He left a last will and testament, naming the executors as Theodore of Reden, Leonard Niederhoff, George Donner and Michael Leutze. The will specified that all of Copernicus' estate was to go to his nieces Catherine and Regina and the seven children of the latter. He left his manuscript of *De revolutionibus* to Tiedemann Giese, who in turn bequeathed it to Rheticus.

It is possible that Anna Schilling may have tried to see Copernicus in his last days, for she was seen in Frauenburg on several occasions after his death. The canons of the Warmia chapter wrote to Bishop Dantiscus on 10 September 1543, asking for advice on the legality of continuing the ban on her presence in town.

> It is not unknown under what circumstances Anna Schilling was banished from here. She used to be the housekeeper of the venerable Doctor Nicolaus, while he was alive. Now from time to time she seems to come here and stay several days for the sake of taking care of her property, as she is said to claim. For she still owns a house here, which she is said to have sold yesterday. We are not sure whether she will be able lawfully to be prohibited from coming here, since the legal obstacle is inoperative. For, when the cause is removed, so is the effect. Nevertheless, we are unable to reach any decision in this matter, unless your Paternal Reverence hands down a prior ruling on this question, since this case was first tried in this court. It will be no trouble to inform us about the decision of your Paternal Presence, whom we commend to God.

Dantiscus wrote back to the chapter immediately, telling the canons that he was not in favour of allowing Anna Schilling to return to Frauenburg, though he left the decision up to them. But he warned them that if they did allow her to stay, she was liable to ensnare one of them just as she seduced their departed brother:

> She, who has been banned from our domain, has betaken herself to you, my brothers. I am not much in favor, whatever the reasons. For it must be feared that by the methods by which she deranged him, who departed from the living a short while ago, she may take hold of another of you, my brothers. But if you have decided to let her stay among your people, that is to be judged by you, my brothers. Nevertheless, I would consider it better to keep at a rather great distance than to let in the contagion of such a disease. How much she has harmed our church is not unknown to you, my brothers, for whom I hope happiness and health.

Copernicus lived almost all of his seventy years in the 'obscure corner of the earth' where he was born, and his death passed unnoticed by all except the few who were devoted to him and the small number of astronomers who were aware of his work. But after his death his reputation began widening, slowly at first and then accelerating, as understanding of his novel theory began spreading through the wider world, starting an intellectual upheaval that came to be known as the Copernican Revolution.

CHAPTER 12
THE REVOLUTIONS OF THE CELESTIAL SPHERES

The title, *On the Revolutions of the Celestial Spheres,* has led some to think that Copernicus believed that the orbital motion of the planets was controlled by the aetherial spheres in which they were supposedly embedded. But it is clear that Copernicus had the planets moving not in spheres but in circular 'orbits,' as indicated in the term '*orbus magnus*,' the 'great circle' known as ecliptic, which for Ptolemy was the circle in which the sun was carried by its celestial sphere around the earth and in the heliocentric theory was the circular orbit of the earth around the sun.

The front papers begin with the *Ad lectorem* and the letter from Cardinal Schönberg, followed by the 'Preface and Dedication to Pope Paul III.' Copernicus begins the preface by telling the pope of why he hesitated for so long before he published *De revolutionibus*:

> I can reckon easily enough, Most Holy Father, that as soon as certain people learn that in these books of mine about the revolutions of the spheres of the world I attribute certain motions to the terrestrial globe, they will immediately shout to have me and my opinion hooted off the stage [...] And when I considered how absurd this 'lecture' would be held by those who know that the opinion that the Earth rests immovable in the middle of the heaven as if their centre had been confirmed

> by the judgments of many ages – if I were to assert to the contrary that the Earth moves, for a long time I was in great difficulty as to whether I should bring to light my commentaries written to demonstrate the Earth's movement, or whether it would not be better to follow the example of the Pythagoreans and certain others who used to hand down the mysteries of their philosophy not in writing but by word of mouth and only to their relatives and friends [...] Therefore, when I weighed these things in my mind, the scorn which I had to fear on account of the newness and absurdity of my opinion almost drove me to abandon a work already undertaken.

He tells the pope of how he was persuaded to change his mind by friends, first by the encouragement of Cardinal Nicolaus Schönberg and then by Bishop Tiedemann Giese, who persuaded him to publish the book. He explains that he was led to begin his work because of his

> knowledge that mathematicians have not agreed with one another in their researches moved me to think out a different scheme of drawing up the movements of the spheres of the world. For in the first place mathematicians are so uncertain about the movements of the sun and moon that they can neither demonstrate nor observe the unchanging magnitude of the revolving year. Then in setting up the solar and lunar movements and those of the other five wandering stars, they do not apply the same principals, assumptions or demonstrations for the revolutions and apparent movements. For some make use of homocentric circles and epicycles, by means of which however they do not attain fully what they seek.

Thus he began to 'reread all the books by philosophers which I could get hold of, to see if any of them even supposed that the movements of the spheres of the world were different from those who taught mathematics in the schools.' He found in reading Cicero and Plutarch references to Pythagoreans who thought that the earth was in motion, which led him to think along the same lines.

> Therefore, I also, having found occasion, began to mediate upon the mobility of the Earth. And although the opinion seemed absurd, nevertheless because I knew that others before me had been granted the liberty of constructing whatever circles they pleased in order to demonstrate astral phenomena, I thought that I too would be readily permitted to test whether or not, by laying down that the earth had some

The Revolutions of the Celestial Spheres

movement, demonstrations less shaky than those of my predecessors could be found for the revolution of the celestial spheres.

He goes on to say that his calculations had shown that the orbits of the other five planets were correlated with that of the earth, so that the solar system was unified and self-consistent.

> And so having laid down the movements which I attribute to the Earth farther on in the work, I finally discovered by the help of long and numerous observations that if the movements of the other wandering stars are correlated with the circular movements of the Earth, and if the movements are computed in accordance with the revolution of each planet, not only do all their phenomena follow from that but also this correlation binds together so closely the order and magnitudes of all the planets and of their spheres or orbital circles and the heavens themselves that nothing can be shifted around in any part of them without disrupting the remaining parts and the universe as a whole

He concludes by pointing out to the Pope that he had also been motivated to help reform the ecclesiastical calendar.

> Mathematics is written for mathematicians; and among, if I am not mistaken, my labours will be seen to contribute something to the ecclesiastical commonwealth, the principate of which Your Holiness now holds. For not many years ago under Leo X when the Lateran Council was considering the question of reforming the Ecclesiastical Calendar, no decision was reached, for the sole reason that the magnitude of the year and the months and the movements and the movements of the sun and moon had not yet been measured with sufficient accuracy. From that time on I gave attention to making more exact observation of these things and was encouraged to do so by that most distinguished man, Paul, Bishop of Fossombrone, who had been present at these deliberations. But what have I accomplished in this matter I leave to the judgment of Your Holiness in particular and to that of all other learned mathematicians. And so as not to appear to Your Holiness to make more promises concerning the utility of this book than I can fulfill, I now pass on to the body of the work.

The first eleven chapters of Book I give a general description of the world and discuss the possible place and motion of the earth. Chapters 1 and 2 present arguments for the spherical shape of the earth. In Chapter 3, on 'How Land and Water Make Up a Single Globe,' Copernicus talks about the recent discovery of the New World and remarks that 'reasons of geometry compel us to believe that America is situated diametrically opposite to the India of the Ganges.' Here he begins to present his defence against physical objections to the earth's rotation, an argument that recurs in several subsequent chapters, starting with his remark that 'it is manifest that the land and water rest upon one centre of gravity; that it is the same as the centre of magnitude of the land, since land is heavier.'

Chapter 4 is entitled 'The Movement of the Celestial Bodies is Regular, Circular, and Everlasting – Or Else Compounded of Circular Movements.' He says that

> The movement of the celestial bodies is circular. For the motion of a sphere is to turn in a circle; by this very act expressing its form [...] But there are many movements on account of the multitude of sphere or orbital circles [...] these movements are circular or are composed of many circular movements, in that they maintain these irregularities in accordance with a constant law and with fixed periodic returns.

Chapter 5, on 'Does the Earth Have a Circular Movement? And of its Place,' explains how the heliocentric theory can be explained by the relativity of motion:

> Although there are so many authorities for saying that the Earth rests in the centre of the world that people think the contrary supposition inopinable and even ridiculous; if however we consider the thing attentively, we will see that the question has not yet been decided and accordingly is by no means to be scorned. For every apparent change in place occurs on account of the thing seen or of the spectator, or on account of the necessarily unequal motion of both. For no movement is perceptible relative to things moved in the same directions – I mean relative to the thing seen and the spectator. Now it is from the Earth that the celestial circuit is beheld and presented to our sight. Therefore, if some movement should belong to the Earth, it will be appear, in the parts of the universe that are outside, as the same movement but in the opposite direction, as though the things outside were passing over. And the daily revolution in especial is such a movement. For the daily revolution appears to carry the whole universe along, with the exception of the Earth and the things around it. And if you admit that the heavens possess none of this movement but that the Earth turns from west to east, you

will find – if you make a serious examination – that as regards the apparent rising and setting of the sun, moon, and stars the case is so.

He goes on to say that the Pythagoreans 'made the Earth to revolve at the centre of the world. For they believe that the stars set by reason of the interposition of the Earth, and with the cessation of that they rose again.' He then examines the consequences of this assumption:

> Now upon this assumption there follow other things, and a no smaller problem concerning the place of the Earth, though it was taken for granted and believed by nearly [all] that the Earth is the centre of the world. For if anyone denies that the Earth occupies the midpoint or centre of the world, yet does not agree that the distance (between the two) is great enough to be compared with (the distance to) the sphere of the fixed stars but is considerable and quite apparent in relation in relation to the orbital circles of the sun and the planets; and if for that reason he thought that their movements appeared irregular because they were organized around a different centre from the centre of the Earth, he might perhaps be able to bring forward a perfectly sound reason for movement which appears regular. For the fact that the wandering stars are seen to be sometimes nearer the Earth and at other times farther away necessarily argues that the centre of the Earth is not the centre of their circles. It is not yet clear whether the Earth draws near to them and moves away or they drawn near to the Earth and move away.

Chapter 6, 'On the Immensity of the Heavens in Relation to the Magnitude of the Earth,' shows that the size of the earth is insignificant compared to that of the universe. As Copernicus argues:

> It can be understood that this great mass which is the Earth is not comparable with the magnitude of the heavens, from the fact that the boundary circles [...] cut the whole celestial sphere into two halves; for that could not take place if the magnitude of the Earth in comparison with the heavens, or its distance from the centre of the world, were considerable. For the circle bisecting a sphere goes through the centre of the sphere, and is the greatest which it is possible to circumscribe.

Chapter 7 is entitled 'Why the Ancients Thought the Earth was at Rest at the Middle of the World at its Centre.' Chapter 8 is an 'Answer to the Aforesaid Reasons and their Inadequacy.'

Chapter 9 is on 'Whether Many Movements can be Attributed to the Earth, and Concerning the Center of the World.' Here Copernicus abandons the Aristotelian doctrine that the earth is the sole source of gravity, and instead takes the first step towards the Newtonian theory of universal gravitation, writing that 'I myself think that gravity or heaviness is nothing except a certain natural appetency implanted in the parts by the divine providence of the universal Artisan, in order that they should unite with one another in their oneness and wholeness and come together in the form of a globe.'

Chapter 10 is entitled 'On the order of the celestial orbital circles.' Here Copernicus removes the ambiguity concerning Mercury and Venus, which in the Ptolemaic model were sometimes placed 'above' the sun and sometimes 'below.' The Copernican system has Mercury as the closest planet to the sun, followed by Venus, earth, Mars, Jupiter and Saturn, surrounded by the sphere of the fixed stars, and with the moon orbiting the earth. This model is simpler and more harmonious than Ptolemy's, for all of the planets revolve in the same sense, with velocities decreasing with their distance from the sun, which sits enthroned at the centre of the cosmos.

> In the center of all rests the Sun. For who would place this lamp of a very beautiful temple in another or better place than this wherefrom it can illuminate everything at the same time. As a matter of fact, not unhappily do some call it the lantern, others the mind and still others, the pilot of the world [...] And so the Sun, as if resting on a kingly throne, governs the family of stars which wheel around.

Chapter 11 is 'A Demonstration of the Threefold Movement of the Earth,' while the remaining three chapters of Book One are concerned with the application of plane and spherical geometry and trigonometry to problems in astronomy. The three motions to which Copernicus refers are the earth's daily rotation on its axis, its yearly revolution around the sun, and a third conical motion, that he introduced to keep the earth's axis pointing in the same direction while the crystalline sphere in which it was embedded rotated annually. The period of this supposed third motion he took to be slightly different than the time it takes the earth to rotate around the sun, the difference being due to the very slow precession of the equinoxes.

Chapters 12–14 are an introduction to spherical trigonometry, with a table of the chords in a circle, which will be used in the books that follow.

Book II is a detailed introduction to astronomy and spherical trigonometry, together with mathematical tables and a catalogue of the celestial coordinates of 1,024 stars, most of them derived from Ptolemy, adjusted for the precession of the equinoxes.

Book III is concerned with the precession of the equinoxes and solstices, which can be understood by referring to the upper drawing in Figure 5. The daily rotation of the earth on its axis and its yearly revolution around the sun define two great circles on the celestial sphere, the two circles making an angle of about 23.5° because of the tilt of the earth's axis. One of these is the celestial equator, the projection of the terrestrial equator on to the celestial sphere. The other is the ecliptic, the projection of the earth's orbit on to the celestial sphere. If the earth's axis of rotation maintained a constant direction relative to the stellar sphere, and if the plane of its orbit were constant as well, then the two equinoctial points, where these two great circles intersect, would remain fixed relative to the stars. But the equinoctial points are not stationary, but move very slowly around the ecliptic relative to the stars, in the opposite direction to the movement of the sun during the year. Thus there is a steady increase in the celestial longitudes of the stars, as Hipparchus discovered. Ptolemy interpreted this as a slow eastward rotation of the celestial sphere about the poles of the ecliptic. The result is that the sidereal year, the time it takes the sun to make to complete one circuit relative to the stars, is greater, by about 20 minutes, than the tropical year, the time for it to complete one revolution relative to the equinoctial points, an effect known as the precession of the equinoxes.

Chapter 1 of Book III begins with a brief historical introduction, in which Copernicus points out that the ancients 'made no distinction between the "turning" or natural year, which begins at an equinox or solstice, and the year which is determined by the fixed stars.' He then writes that

> Hipparchus [...] was the first to call attention to the fact that there was a difference in the length of these two kinds of year. While making careful observations of the magnitude of the year, he found that it was longer

as measured from the fixed stars than as measured from the equinoxes or solstices. Hence he believed that the fixed stars too possessed a movement eastward, but one so slow as not to be immediately perceptible.

Copernicus rejects the idea that the effect is due to a motion of the stars, arguing that it is caused by a continuous change in the plane of the earth's equator, in which the earth's axis of rotation describes a cone about the perpendicular to the ecliptic. He explains that this is due to the fact that there is a slight difference between two of the three motions he attributes to the earth, namely an annual revolution from west to east in the plane of the ecliptic, and a variation of the earth's axis to the line joining sun and earth, of slightly smaller period as the former, but in the opposite sense, with the obliquity maintaining a fairly constant value.

> For [...] the two revolutions, that is of the annual declination and of the centre of the Earth, are not altogether equal, namely because the restoration of the declination anticipates the period of the centre, whence it necessarily follows that the equinoxes seem to arrive before their time – not that the sphere of the fixed stars is moved eastward, but rather that the equator is moved westward, as it is inclined obliquely to the plane of the ecliptic in proportion to the amount of deflexion of the axis of the terrestrial globe.

Copernicus' theory is unnecessarily complicated, since he found it compulsory to account for five illusory secondary movements of long periods that ancient astronomers seem to have observed. These were: a variation in the length of the tropical year; a variation in the rate of precession; a decrease of the obliquity of the ecliptic; a decrease in the departure of the sun from uniform angular motion; a non-uniform motion of the apsidal line of the sun (or of the earth). (The apsidal line is the line connecting the earth, the centre of the eccentric orbit of the sun (or of the earth), and the apogee and perigee, the points where the distance between the earth and the sun are a maximum or minimum, respectively.)

The last subject dealt with in Book III, Chapter 26, is what Copernicus calls the 'correction of the natural day.' This is equivalent to the modern 'equation of time,' the difference between the right ascension of the sun and of a fictitious mean sun that moves along, or parallel to, the equator. (The right ascension is the celestial longitude of a celestial body, measured eastward along the celestial equator from the spring equinox, while declination is its celestial latitude, its angular distance north or south of the plane of the celestial equator.)

Book IV deals with the motion of the moon about the earth. The moon's orbit is inclined by about 5° to the ecliptic. At full moon the moon is usually a few degrees above or below the ecliptic, and is thus outside the earth's shadow cone. Lunar eclipses occur only when the sun is at or near one of the two nodes of the lunar orbit, the points where the orbit passes through the plane of the ecliptic. Lunar eclipses occur only about twice a year, when the sun and moon are simultaneously at opposite nodes. The nodes of the lunar orbit move westward around the ecliptic in a period of about 18.6 years.

These lunar periodicities are called the synodic, tropical, anomalistic and draconitic months. A synodic month (29.5306 days) is the mean time between successive full moons. A tropical month (27.3216 days) is the mean time for the moon to travel around the ecliptic from an equinoctial point back to the same equinoctial point. An anomalistic month (27.5546 days) is the mean time for the moon to travel around the ecliptic from the perigee, the point where it is closest to the earth, back to the perigee. The draconitic month (27.2122 days) is the time it takes the moon to travel around the ecliptic from a node back to the same node.

The latter name comes from the constellation Draco, or the Dragon. Copernicus, explaining the lunar periodicities in *De revolutionibus*, IV, 10, writes of the moon being near 'the mean position between the southern limit of latitude and the ascending ecliptic section – which the moderns call the head of the Dragon – and the sun had already passed by the other descending section – which they call the tail.'

The lunar inequalities, or anomalies, are periodic fluctuations in the moon's rate of motion about the earth. Only two inequalities were known to Ptolemy, and no more were discovered until after the time of Copernicus. The first inequality is an oscillation of the moon about the mean position in its orbit, due to the eccentricity of the lunar orbit. The second inequality is a variation in the first one, caused by fluctuations in the eccentricity of the lunar orbit.

The subjects dealt with by Copernicus in Book IV involve the first and second inequalities of the moon; the parallax, distance and size of the sun and moon; the apparent diameters of the sun, moon, and shadow cone of the earth; the theory of eclipses; and the lunar periodicities represented by the synodic, tropical, anomalistic and draconitic months.

Chapters 1 and 2 of Book IV deal with problems in Ptolemy's lunar model, while in Chapter 3 Copernicus presents his own model. In both models the moon orbits the earth, but the Copernican model attributes the rising and setting of the moon to the rotation of the earth on its axis.

Copernicus raised four objections to the Ptolemaic model. These can be summarized by saying that it did not preserve the principle of uniform circular motion, and that it gave a variation in the lunar distance incompatible with the observed variation of the moon's parallax and apparent diameter. Thus, as Copernicus says in the introduction to Book IV:

> In our explanation of the circular movement of the moon we do not differ from the ancients as regards the opinion that it takes place around the Earth. But we shall bring forward certain things which are different from what we received from our elders and are more consonant; by means of them we shall try to set up the movement of the moon with more certitude, in so far as that is possible.

The lunar model proposed by Copernicus was anticipated by that of the Arabic astronomer Ibn al-Shatir (d. *c.* 1375), whom he does not mention in *De revolutionibus*. Neither does he mention three other Arabic astronomers whose works anticipated other models used by Copernicus: Mu'ayyad al-Din al-'Urdi (d. 1266), Nasir al-Din al-Tusi (d. 1274) and Qutb al-Din al-Shirazi (d. 1312). These four astronomers were members of the Maragha School, which takes its name from the observatory built in 1259 at Maragha in northwestern Iran during the reign of the Ilkhanid Mongols. Swerdlow and Neugebauer write of these astronomers in their *Mathematical Astronomy in Copernicus' De revolutionibus*:

> The planetary models for longitude in the *Commentariolus* are all based upon the models of Ibn al-Shatir – although the arrangement for the interior planets is incorrect – while those for the superior planets in *De revolutionibus* use the same arrangements as 'Urdi's and Shirazi's model, and for the inferior planets the smaller epicycle is converted into an equivalent rotating eccentricity that constitutes a correct adaption of

Ibn al-Shatir's model. In both the *Commentariolus* and *De revolutionibus* the lunar model is identical to Ibn al-Shatir's and finally in both works Copernicus makes it clear that he is addressing the same physical problems as his predecessors. It is obvious that with regard to these problems his solutions were the same. The question therefore is not whether, but when, where, and in what form he learned of Maragha theory.

Swerdlow has reinforced this conclusion in a more recent publication, where he writes that: 'How Copernicus learned of the models of his [Arabic] predecessors is not known – a transmission through Italy is the most likely – but the relation between the models is so close that independent invention by Copernicus is all but impossible.'

The Copernican lunar model is shown in the middle drawing in Figure 5, where E is the earth, AB is a large epicycle with its centre at C, and DF is a smaller epicycle with its centre at A. Starting from F, the moon moves from west to east, while A moves in the opposite sense and C moves from west to east. When EC is in the direction of the mean sun, the moon is at D, and when EC has rotated 90° from this position the moon is at F. It thus goes around the circle DF twice in a synodic month relative to the line DC. At the same time A traverses the larger epicycle in an anomalistic month, while C makes a circuit of its deferent (relative to the mean sun) in one synodic month.

Chapter 4 has three sets of tables giving the day-to-day and year-to-year movement of the moon. The first gives the accumulated angular movement relative to the sun (synodic motion); the second relative to the apses of the lunar orbit (anomalistic motion; and the third relative to its nodes on the ecliptic (draconitic motion).

Chapters 5–12 deal with the lunar inequalities, and are represented by the two epicycles in the lower diagram of Figure 5. The problem was to determine the elements of his lunar theory by using observations to determine the ratios (CD:EC) and (CF:EC). According to Swerdlow and Neugebauer:

> Copernicus gives a total of 14 lunar observations. Eight of the observations were his own, and six were from the *Almagest*. Most of the parameters of his lunar theory are derived from the eclipses, and of the five remaining observations, one is used as a test of the lunar theory, two for the determination of lunar parallax, one is a test of parallax, and one that is not used shows Ptolemy's lunar parallax.

Chapter 13 is 'On the Positions of Lunar Anomaly in Latitude,' and Chapter 14 is 'On the Positions of Lunar Anomaly in Latitude.' Chapter 13 is devoted to the evaluation of the draconitic month. Copernicus did this by comparing two partial lunar eclipses occurring at two nearly diametrically opposite points of the moon's orbit relative to the line of nodes. One of these eclipses was observed in 174 BC by Ptolemy and the other in 1509 by Copernicus.

Chapters 15–27 deal with topics required primarily for the calculation of eclipses, principally that of lunar parallax. The main objection that Copernicus had to Ptolemy's lunar model was that it would greatly exaggerate the apparent diameter and parallax of the moon near quadrature. This is when a triangle formed by the earth, sun and moon forms a right angle at the moon, which occurs at half-moon. It is obvious that the apparent lunar diameter does not increase perceptibly near quadrature. Copernicus demonstrated that near quadrature the parallax is less than predicted by the Ptolemaic model. He did this by two direct measurements of the lunar parallax very close to quadrature, the first on 27 September 1522 and the second on 7 August 1524, both in Frauenburg, and by a computation of an occultation of Aldebaran by the moon, the observation that he made in Bologna with Domenico Maria da Novara on 9 March 1497. According to Swerdlow and Neugebauer, the latter observation 'is without doubt the most remarkable demonstration in all of *De revolutionibus.*' They go on to say that 'The problem of the occultation is such that it constitutes a test of nearly everything that has preceded it – spherical astronomy, star catalogue, precession, solar theory, lunar theory, lunar parallax – and the results are by any standard extraordinary.'

Chapters 28–32 are concerned with the theory of eclipses. In Chapter 28 Copernicus computes the time of mean conjunction and opposition for a syzygy, meaning, either a solar or lunar eclipse, which would be conjunction for a solar eclipse and an opposition for a lunar eclipse. In Chapter 29 he finds the time of true syzygy.

Chapter 30 is entitled 'How the Ecliptic Conjunctions and Oppositions of the Sun and Moon are distinguished from the Others.' Copernicus explains what he means by this:

> In the case of the moon it is easily discernable whether or not they are ecliptic; since, if the latitude of the moon is less than half the diameters of the moon and the shadow, it will undergo an eclipse, but if greater,

> it will not. But there is more than enough bother in the case of the sun, as the parallax of each of them, by which for the most part the visible conjunction differs from the true, is mixed up in it. Accordingly when we have examined what the parallax in longitude between the sun and the moon at the time of true conjunction, similarly we shall look for the apparent (angular) elongation of the sun from the moon at the interval of an hour before the true conjunction in the eastern quarter of the ecliptic or after the true conjunction in the western quarter, in order to understand how far the moon seems to move away from the sun in one hour.

He goes on to describe the calculations that must then be done in order to determine 'the time of true conjunction we are looking for.'

Chapter 31 is entitled 'How Great an Eclipse of the Sun or Moon Will Be', and that of Chapter 32 is 'How to Know Beforehand How Long an Eclipse Will Last.' After Copernicus examines these questions in detail, he concludes Book IV by saying that 'Let all of this – which has been treated in detail by others – be enough now concerning the moon: for we are in a hurry to get to the revolutions of the remaining five planets, which will be spoken of in the books following.'

Book V concerns the planetary theory of longitude and Book VI the planetary theory of latitude. In the introduction to Book V Copernicus states his aim in writing these two books, the last in *De revolutionibus*:

> Up to now we have been explaining to the best of our ability the revolutions of the Earth around the sun and of the moon around the Earth. Now we are turning to the movements of the five wandering stars: the mobility of the Earth binds together the order and magnitude of their orbital circles in a wonderful harmony and sure commensurability, as we said in our brief survey in the first book, when we showed that the orbital circles do not have their centres around the earth but rather around the sun. Accordingly it remains for us to demonstrate these things singly and with greater clarity: and let us fulfill our promises adequately, insofar as we can, particularly by measuring the appearances by the experiments which we have got from the ancients or from our own times, in order that the ratio of the movements may be held with greater certainty

Chapter 1 of Book V describes the revolutions and mean movements of the planets. Here Copernicus distinguishes between the two components in the apparent motion of a planet in longitude. The first is its orbital motion around the sun, completed in a sidereal year, meaning, as it would be measured from the sun against the background of the fixed stars. The second is the apparent or parallactic motion, which superimposes upon its orbital motion an inequality which recurs in the planet's synodic period, in other words, as observed from the earth, causing an apparent periodic retrograde motion.

When observed from the earth, the sun and the superior planets exhibit two so-called anomalies or inequalities. The first, or zodiacal anomaly, is due to the fact that the sun appears to move faster in some parts of the zodiac and more slowly in others. The second, or solar anomaly, is the apparent retrograde motion that the superior planets exhibit when they are in opposition to the sun, for instance, when they are in opposite directions from the earth.

At the beginning of Chapter 1 Copernicus describes these two anomalies for the superior planets:

> Two longitudinal movements which are quite different appear in the planets. One of them is on account of the movements of the Earth, as we said; and the other is proper to each planet. We might rightly call the first the movement of parallax, since it is the one which makes the planets appear to have stoppings, progressions, and retrogradations – not that the planet which always progresses by its own movement is pulled in different directions, but that it appears to do so by reason of the parallax caused by the movement of the Earth taken in relation to the differing magnitudes of their orbital circles. Accordingly it is clear that the true position of Saturn, Jupiter and Mars becomes visible to us only when they are in opposition to the sun; and that occurs approximately in the middle of their retrogradations. For at that time they fall on a straight line with the mean position of the sun, and lay aside their parallax.

Copernicus then goes on to discuss the interior planets: 'Furthermore there is a different ratio in the case of Venus and Mercury; for they are hidden at the time they are in conjunction with the sun, and they show only the digressions which they make on either side away from the sun: and hence they are never found without parallax.'

Chapter 2 deals with the first, or zodiacal anomaly, while Chapter 3 is concerned with second, or solar anomaly.

The basic planetary model used by the ancient Greeks is shown in Figure 3. In the upper drawing the planet P revolves in a small circle, the

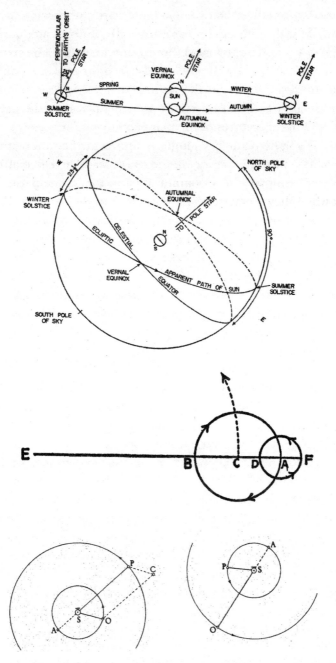

Figure 5 The precession of the equinoxes (above); Copernican lunar model (middle); Copernican model for the solar anomaly (below) of a superior planet (left) and an inferior planet (right).

epicycle, whose centre C moves in a larger circle, the deferent, around the earth E, with both circles revolving with uniform angular velocity from west to east (counter-clockwise in the drawing). The period of P is the planet's mean synodic period. By giving the proper ratio of the radii of the two circles and the periods of their motion it is possible to reproduce a reasonably accurate representation of the retrograde motion and other features of the planetary orbits.

Ptolemy's basic planetary model is shown in the lower drawing of Figure 3. The planet P moves with constant angular velocity from west to east on an epicycle whose centre C moves in the same sense on a deferent with its centre at O, which is eccentric to the earth, E, The point C moves with constant angular velocity with respect to the point O', where EO'= 2EO, where O' is the so-called equant point.

Ptolemy used this model to describe the motions of Saturn, Jupiter, Mars and Venus. The motion of Mercury is much more complex, and more elaborate models were needed to describe its motion. There were also elaborate models for representing the planetary motions in latitude.

Copernicus particularly objected to Ptolemy's use of the equant, which he felt violated the principle that the celestial motions should be circular with uniform angular velocity. As he says at the end of Chapter 2: 'These and similar things furnished us with an occasion for working out the mobility of the earth and some other ways by which regularity and the principles of this art might be preserved, and the ratio of apparent irregularity rendered more constant.'

As Swerdlow and Neugebauer point out, 'Copernicus's most well-known contribution to astronomy is his model for the second or solar anomaly of the planets'. Their description of his model is illustrated by the lower drawing in Figure 5, for superior planets on the left and inferior planets on the right, where S is the mean sun, O the earth, and P, the planet, moving in a epicycle with centre C, the motion of both C and P being in the counterclockwise sense, and A the apogee of the epicycle. They first describe the model for a superior planet

> The period of P is the planet's mean synodic period, return to conjunction with the mean sun [...] at which time the planet is at the apogee A of the epicycle, and thus the radius of the epicycle is always parallel to the direction [...] from the earth to the mean sun. It follows also that when at opposition to the mean sun the planet will be at the perigee of the epicycle, and that the apparent retrograde motion will take place

in the vicinity of the perigee [...] OS may be taken to be the common measure of the radii SP of the planetary orbits, thereby fixing the order and distances of the planets with respect to the mean sun.

They then describe the model for an inferior planet, noting that 'the center of the epicycle S always in the direction [...] from the earth to the mean sun while the planet P completes a rotation on the epicycle in its mean synodic period. At superior conjunction the planet is at the apogee A of the epicycle.' (An inferior planet is in superior conjunction when the sun is between the earth and the planet; inferior conjunction is when an inferior planet is between the earth and the sun.)

They go on to say that again 'the distance SP may be measured in units of OS, thus determining the distances of the planets from the mean sun.'

Chapter 4 presents the Copernican model for the first anomaly. As mentioned earlier, the planetary models for longitude for the superior planets in *De revolutionibus* use the same arrangements as the models of al-'Urdi and al-Shirazi, and for the inferior planets the smaller epicycle is converted into an equivalent rotating eccentricity that constitutes a correct adaption of Ibn al-Shatir's model.

According to Swerdlow and Neugebauer, the Ptolemaic and Copernican models differ only slightly, aside from the fact that the first is geocentric and the second heliocentric, as can be seen by comparing the two models:

For Ptolemy's model:

(1) The centre of the epicycle describes a circle with its centre removed from the earth by an eccentricity e.
(2) The line joining the centre of the epicycle to a point on the apsidal line removed from the earth by a double eccentricity 2e describes an angle proportional to time.

For Copernicus's model:

(1) The planet describes a figure differing from a circle with its centre removed from the mean sun by an eccentricity e.
(2) The line joining the planet to a point on the apsidal line removed from the mean sun by a double eccentricity 2e describes an angle proportional to time.

Chapters 5–32 deal with the individual planets: Saturn in 5–9, Jupiter in 10–14, Mars in 15–19, Venus in 20–24, and Mercury in 25–32. Chapters 33 and 34 describe the planetary tables and their use, while the last two chapters, 35 and 36, deal with Apollonius' theorem to explain retrograde motion.

The mean distances of the planets from the sun determined by Copernicus, taking the radius of the earth's orbit as 1.0000, are close to the true values, as shown below, where the true values are in parentheses:

Mercury = 0.3763 (0.3871)
Venus = 0.7193 (0.7233)
Earth = 1.0000 (1.0000)
Mars = 1.5198 (1.5237)
Jupiter = 5.2192 (5.2028)
Saturn = 9.1743 (9.5389)

As noted previously, Book VI, the last and shortest book of *De revolutionibus*, deals with the problem of giving a geometric demonstration of the deviation of planetary orbits from the ecliptic plane.

The introduction to Book VI states Copernicus' purpose:

> We have indicated to the best of our ability what power and effect the assumption of the revolution of the earth has in the case of the apparent movement in longitude of the wandering stars and in what a sure and necessary order it places all the appearances. It remains for us to occupy ourselves with the movements of the planets by which they digress in latitude and to show how in this case too the self same mobility of the Earth exercises its command and prescribes laws for them here also. Moreover this is a necessary part of the science, as the digressions of these planets cause no little variation in their rising and setting, apparitions and occultations, and the other appearances of which there has been a general exposition above. And their true positions are said to be known only when their longitude together with their latitude in relation to the ecliptic has been established. Accordingly by means of the assumption of the mobility of the Earth we shall do with perhaps greater compactness and more becomingly what the ancient mathematicians thought to have demonstrated by means of the immobility of the Earth.

Copernicus' latitude theory, which deals first with the superior planets and then the inferior, is based on that of Ptolemy, not only in the structure of his models but also in the data that he used to derive his parameters. The astronomer Johannes Kepler, writing early in the seventeenth century, was the first to point this out, saying that 'Copernicus, ignorant of his own riches, took it upon himself for the most part to represent Ptolemy, rather than nature, to which he had nevertheless come the closest of all.'

Swerdlow and Neugebauer, in quoting Kepler's statement, go on to say that he was referring 'specifically to the "librations" of the inclinations of the planes of the eccentrics, not in accordance with the motion of the planet, but [...] the unrelated motion of the earth.' They then say that this 'was the result of Copernicus's attempt to duplicate the apparent latitudes of Ptolemy's models in which the inclinations of the epicyclic planes were variable.' They go on in more detail:

> But Copernicus's representation, or better, imitation of Ptolemy in Book VI goes farther than the structure of his models. The 'observational' basis of Copernicus's parameters was neither his own observations nor even Ptolemy's, but the extremal latitudes in Ptolemy's tables. And when he found, in the case of Mars, that his model could not reproduce the extrema of the tables, he changed the 'observations' to the closest values that his model would fit, or at least he attempted with only partial success to do so. Still more remarkable, even though his model for the latitude of the inferior planets is equivalent to Ptolemy's only when the earth is in the apsidal line or 90° from the apsidal line, Copernicus nevertheless copied out Ptolemy's tables and applied them indifferently to all locations of the earth, although neither the tables nor the method of computing latitudes from them are applicable to his model.

Chapters 1–4 cover the planetary theory of latitude for the superior planets, with the inferior planets treated in Chapters 5–8 as well as in Chapters 1–2, where the Copernican model of latitude for all the planets is developed. The inclination of the orbital planes of the superior planets to the ecliptic plane is relatively small: for Saturn 2.49°, for Jupiter 1.31° and for Mars 1.85°. Those of the inferior planes are significantly larger: 3.39° for Venus and 7.00° for Mercury. The values given by Copernicus were: Saturn = 2°30', Saturn = 1°30', Mars = 1°0', Venus = 2°30', Mercury = 6°30'.

The Copernican model for the motions of the superior planets in latitude is described in Chapters 1–2. The maximum northward and southward displacements of each planet are at diametrically opposite points in its orbit. Midway between these limits the planet passes through the ecliptic at its nodes. The Copernican model has the line of nodes passing through the centre of the earth's eccentric orbit rather than through the sun. The periodic changes in the planet's latitude also depend on the planet's position relative to the earth, partly due to variations in the distance between the two bodies, though Copernicus attributed it partly to a periodic variation in the planetary orbit's plane to that of the ecliptic.

Chapter 3 gives the mean inclinations of the orbits of the superior planets as well as the amplitudes of their oscillations, based on observations of the maximum angular inclination of the planets opposition and at conjunction. Chapter 4 explains how to calculate the apparent latitude of a superior planet from any point in the earth's orbit.

As mentioned earlier, Chapters 5–8 describe the Copernican theory of latitude for the inferior planets. The orbit of each inferior planet is intersected by the ecliptic in its apse line, and the greatest latitude should occur when the planet is 90° from apogee, but, according to Copernicus, there are two types of so-called librations. The first has a periodicity of half a year, so that whenever the mean sun passes through the planet's apogee or perigee the inclination is greatest. The second libration, which produces the greatest deviation, occurs around a moving axis, with the planets passing through it whenever the earth is 90° from the apsides. When the apogee or perigee of the planet is turned towards the earth, Venus always has its maximum deviation to the north and Mercury to the south, with the periodicity equal to a year for both planets.

Chapter 9, the last in Book VI entitled 'On the Calculation of the Latitudes of the Five Wandering Stars', explains the use of the tables giving the latitudes of the five planets.

Swerdlow and Neugebauer give this critical appraisal of Book VI in their concluding remarks:

> The latitude theory of the planets is certainly the most flawed part of *De revolutionibus*. Rather than constructing his theory upon independent observations – which were in any case probably unobtainable in the required locations of the earth and each planet – Copernicus set the goal of producing as directly as possible in a heliocentric form the latitudes of Ptolemy's tables in the *Almagest*. But even this more modest task was beyond Copernicus's reach, and led to a series of problems,

the most notable being the inexplicable relation of the oscillating planes of the planets to the motion of the earth, the violation of uniform circular motion, and finally the incompatibility of the model for the inferior planets with the very tables they were intended.

They conclude their book with the statement that 'Book VI must have been revised into its present form during the period that Copernicus was preparing *De revolutionibus* for publication, and in his haste he may have been less than careful about the undesirable consequences of his attempt to represent Ptolemy rather than nature.'

Thus it would seem that Book VI took on its final form in the spring and summer of 1541, just before Copernicus gave Rheticus the final version of *De revolutionibus* to be taken to the printer.

CHAPTER 13
THE COPERNICAN REVOLUTION

Two ephemerides, or almanacs giving the positions of the celestial bodies for the coming year, were published in 1550, both of them based on the Copernican theory. One of them was by Rheticus, who at Leipzig published 5,000 copies of an almanac giving the positions of the celestial bodies for the year 1551. The title of the work was *Ephemerides: A Setting Forth of the Daily Position of the Stars [...] by Georg Joachim Rheticus according to the theory [...] of his teacher Nicolaus Copernicus of Torun*. These were computed directly from the tables in *De revolutionibus*, as Rheticus notes in his statement that 'I have not wanted to backslide from Copernican teaching, not even by a finger's width.'

The other astronomical almanac was published in Tübingen by Erasmus Reinhold (1511–53), Professor of Higher Mathematics (i.e., astronomy) at the University of Wittenberg at the same time as Rheticus. His ephemerides for the years 1550 and 1551 were calculated from his *Prutenicae tabulae* (Prutenic Tables), published in 1551, so-called in honour of Reinhold's patron, Duke Albrecht of Prussia. Reinhold, in the dedication to the *Prutenicae tabulae*, refers to Copernicus as 'the most learned man whom we may call a second Atlas or a second Ptolemy.' Nevertheless Reinhold criticized the tables in *De revolutionibus*, stating that 'the computation is not even in agreement with his observations on which the foundation of his work rests.' But Owen Gingerich, in his study comparing the two sets of tables, concludes that 'Although it is clear in many cases that Reinhold was the more fastidious calculator, he was deluding himself if he thought he had made any significant improvement over Copernicus's prediction of planetary positions.'

The second printing of *De revolutionibus* in Basel in 1566 adds this encomium by Reinhold: 'All posterity will gratefully remember the name of Copernicus, by whose labor and study the doctrine of celestial motions was again restored from its near collapse. Under the light kindled in him by a beneficent God, he found and explained much which from antiquity till now was either unknown or veiled in darkness.'

The *Prutenic Tables* were much easier to use than those in *De revolutionibus*, which has the tables scattered through the text. Reinhold, in one of the two prefaces to the *Prutenic Tables*, calls them 'Very convenient tables from which you can most easily calculate (the positions of the planets) for any time.' But although the *Prutenic Tables* were based on *De revolutionibus*, Reinhold did not say whether or not the heliocentric theory of Copernicus was physically true. The *Prutenic Tables* were the first complete planetary tables prepared in Europe since the *Alfonsine Tables*. They were demonstrably superior to the older tables, which were now out of date, and so they were used by most astronomers, lending legitimacy to the Copernican theory even when those who used them did not acknowledge the sun-centred cosmology of Copernicus. As the English astronomer Thomas Blundeville wrote in the preface to an astronomy text in 1594: 'Copernicus [...] affirmeth that the earth turneth about and that the sun standeth still in the midst of the heavens, by help of which false supposition he hath made truer demonstrations of the motions and revolutions of the celestial spheres, than ever were made before.'

The English mathematician Robert Recorde (1510–58) was one of the first to lend some support to the Copernican theory. He discusses the theory in his *Castle of Knowledge* (1551), an elementary text on Ptolemaic astronomy with a brief favourable reference to the Copernican theory. It is written in the form of a dialogue between a Master and a Scholar concerning Ptolemy's arguments against the earth's motion. After the Scholar sums up these arguments, the Master presents the Copernican theory in a very positive manner.

> That is trulye to be gathered: howe be it, COPERNICUS, a man of greate learning, of much experience, and of wonderfull diligence in observation, hath renewed the opinion of ARISTARCHUS SAMIUS,

and affirmith that the earthe not only moveth circularlye about his own centre, but also may be, yea and is, continually out of the precise centre of the world 38 hundred thousand miles.

Five years later an *Ephemeris* was printed in London for the year 1557 by John Feild, at the request of John Dee (1527–1608), Queen Elizabeth's astrologer. Feild notes that he had written his *Ephemeris* because the errors in the *Alfonsine Tables* were becoming more apparent every day. He goes on to say that 'Wherefore, I have published this *Ephemeris* for the year 1557, following in it Copernicus and Erasmus Reinhold, whose writings are established and founded on true, sure and plain demonstrations.' Dee, in an epistle attached to the *Ephemeris*, praised Copernicus for his 'more than Herculean' effort in restoring astronomy, though he said that this was not the place to discuss the heliocentric theory itself.

The English astronomer Thomas Digges (*c.*1546–95), a pupil of Dee, obtained a copy of *De revolutionibus*, the Basel edition of 1566, which has survived in the library of Geneva University, along with a note he wrote on the title page, *'Vulgi opinio error'* (the common opinion errs), indicating that he was one of the few sixteenth-century scholars who accepted the Copernican theory.

Digges did a free English translation of Chapters 9 through 11 of the first book of *De revolutionibus*, adding it to his father's perpetual almanac, *A Prognostication Everlasting*, publishing them together in 1576 as a *Perfit Description of the Caelestiall Orbes, according to the most ancient doctrines of the Pythagoreans lately revived by Copernicus and by Geometricall Demonstrations approved*. Digges stated that he had included this excerpt from *De revolutionibus* in the almanac 'so that Englishmen might not be deprived of so noble a theory.'

Digges published a new edition of the *Prognostication Everlasting* in 1592. In an appendix, after describing the Ptolemaic system, he writes, obviously referring to Copernicus, that 'In this our age, one rare witte [...] hath by long study, paynfull practice, and rare invention, delivered a new Theorick or Model of the world.'

The book was accompanied by a large folded map of the sun-centred universe, in which the stars were not confined to the outermost celestial sphere but scattered outward indefinitely in all directions. Digges thus burst the bounds of the medieval cosmos, which till then had been limited by the ninth celestial sphere, the one containing the supposedly fixed stars, which in his model extended to infinity.

The concept of an infinite universe was one of the revolutionary ideas for which the Italian mystic Giordano Bruno (1548–1600) was condemned by the Catholic Church, which had him burned at the stake in Rome on 17 February 1600. Bruno expressed his belief in the Copernican theory on several occasions, most notably in a lecture he gave at Oxford in 1583.

At the beginning of his dialogue on *The Infinite Universe and the Worlds*, published in 1584, Bruno says, though one of his characters, that in this limitless space there are innumerable worlds similar to our earth, each of them revolving around its own star-sun. 'There are then innumerable suns, and an infinite number of earths revolve around these suns, just as the seven [the five visible planets plus the Earth and its Moon] we can observe revolve around this Sun which is close to us.'

The concept of an infinite universe appears also in the work of the English scientist William Gilbert (1544–1603), who may have been influenced in this regard by Thomas Digges and Giordano Bruno. Gilbert's *De Magnete*, published in 1600, was the first work on magnetism since that of Petrus Peregrinus in the thirteenth century.

The sixth and final book of *De Magnete* was devoted to Gilbert's cosmological theories, in which he rejected the crystalline celestial spheres of Aristotle and said that the apparent diurnal rotation of the stars was actually due to the axial rotation of the earth, which he believed to be a huge magnet. His rejection of diurnal stellar motion was due to his belief that the stars were limitless in number and extended to infinity, so that it was ridiculous to think that they rotated nightly around the celestial pole. While Gilbert discussed the motions of the earth according to Copernicus and Giordano Bruno, he did not affirm or deny the heliocentric theory; at times he dismissed it as not being pertinent to the topic he was discussing.

The Italian scientist Giovanni Battista Benedetti (1530–90), writing in 1585, refuted a number of Aristotelian theories of motion. He also rejected the geocentric model of the universe and said that he preferred the 'theory of Aristarchus, explained in a divine manner by Copernicus, against which the arguments of Aristotle are of no value.' He even suggested that the other planets were inhabited, since it was unlikely that the earth was the only site of creation.

Meanwhile astronomy was being revolutionized by the Danish astronomer Tycho Brahe (1546–1601), who in the last quarter of the seventeenth century made systematic observations of significantly greater accuracy than any ever done in the past, all just before the invention of the telescope.

Tycho, born to a noble Danish family, was brought up by his paternal uncle, Jörgen Brahe, who hired a tutor to teach him Latin and the preparatory subjects for a university education. He enrolled at the University of Copenhagen at the age of 13, and subsequently continued his studies at the universities of Leipzig, Rostock and Basel, from which he graduated in 1568. The astronomy books that he studied included Sacrobosco's *De Sphaera*, which had been in use since the thirteenth century, and other texts based on the homocentric spheres of Aristotle and the epicycles and eccentrics of Ptolemy.

The occurrence of a solar eclipse on 21 August 1560, though only partial at Copenhagen, aroused Tycho's interest in observational astronomy. He immediately bought a set of ephemerides based on Reinhold's *Prutenic Tables*, which correctly predicted the date of the eclipse. This deeply impressed him, and he later wrote that it was 'something divine that men could know the motion of stars so accurately that they could long before foretell their places in relative positions.' Through this ephemerides he first became aware of the work of Copernicus, whom he called 'a second Ptolemy.'

Tycho made an observation at Leipzig in August 1563 which he considered to be a turning-point in his career, when he noted a conjunction of Saturn and Jupiter, measuring the angular distance between the two planets using a pair of compasses. He found that the *Alfonsine Tables* were a month off in predicting the date of the conjunction, and that the ephemerides based on the *Prutenic Tables* were several days in error. Thus he began making observations with a home-made instrument known as a radius or cross-staff, and since it was not very accurate, he devised a table of corrections to make up for its deficiencies.

Early in 1569 Tycho went to Augsburg, in Germany, where he made his first observation on 14 April. The instruments that he designed and built for his observations included a great quadrant with a radius of some 19 feet for measuring the altitude of celestial bodies. He also constructed a huge sextant with a radius of 14 feet for measuring angular separations, as well as a celestial globe 10 feet in diameter on which to mark the positions of the stars in the celestial map that he began to create.

Tycho returned to Denmark in 1570, and after his father's death on 9 May of the following year he moved to Herrevad Abbey, the home of his maternal uncle, Steen Bille. On 11 November 1572, as Tycho was walking back to supper from his laboratory, he saw a nova, or new star, that had suddenly appeared in the constellation Cassiopeia, exceeding even the planet Venus in its brilliance. (We now know that this was a supernova, caused by the internal collapse of a supergiant star at the end of its life, releasing an enormous amount of energy.) As Tycho later wrote in his tract on the nova: 'I knew perfectly well – for from my earliest youth I have known all the stars in the sky, something which one can learn without difficulty – that no star had ever existed in that place in the heaven, not even the very tiniest, to say nothing of a star of such striking clarity.'

Tycho's measurements indicated that the nova was well beyond the sphere of Saturn, and the fact that its position did not change showed that it was not a comet. This was clear evidence of a change taking place in the celestial region, where, according to Aristotle's doctrine, everything was perfect and immutable.

The nova eventually began to fade, its colour changing from white to yellow and then red, finally disappearing from view in March 1574. By then Tycho had written a brief tract entitled *De nova stella* (The New Star), which was published at Copenhagen in May 1573. After presenting the measurements that led him to conclude that the new star was in the heavens beyond the planetary spheres,

Tycho expressed his amazement at what he had observed. 'I doubted no longer,' he wrote. 'In truth, it was the greatest wonder that has ever shown itself in the whole of nature since the beginning of the world, or in any case as great as [when the] Sun was stopped by Joshua's prayers.'

The tract impressed King Frederick II of Denmark, who gave Tycho an annuity along with the small offshore island of Hveen, in the Oresund Strait north of Copenhagen, the revenues of which would enable him to build and equip an observatory. Tycho settled on Hveen in 1576, calling the observatory Uraniborg, meaning 'City of the Heavens.' The astronomical instruments and other equipment of what came to be a large research centre were so numerous that Tycho was forced to build an annex called Stjernborg, 'City of the Stars,' with subterranean chambers to shield the apparatus and researchers from the elements. That same year Tycho and his assistants began a series of observations of unprecedented accuracy and precision that would continue for the next two decades.

Tycho's main project at Uraniborg was to make new and more accurate determinations of the celestial coordinates of the fixed stars, and to observe the changing positions of the sun, moon and planets for the purpose of improving the theories of their motions. Tycho produced a catalogue giving the coordinates of 777 fixed stars, to which he later added another 223 so as to bring the total to 1,000.

The celestial coordinates given in Tycho's star catalogue had a mean error, compared to modern values, of less than 40 seconds of arc, far less than that of any of his predecessors. Comparing the coordinates of the 21 principal stars in his catalogue with those measured from antiquity up to his own time, Tycho computed a value for the rate of precession of the equinoxes equal to 51 seconds of arc per year, as compared to the modern value of 50.23 seconds. He correctly assumed the precession to be uniform, making no mention of the erroneous trepidation theory that had caused unnecessary problems for Copernicus.

Shortly after sunset on 13 November 1577 Tycho first noticed a spectacular comet with a very long tail, and he continued to observe it nightly until 26 January of the following year, by which time it had faded to the point that it was hardy visible. His detailed observations of the comet led him to conclude that at its closest approach it was farther away than the moon, in fact even beyond the orbit of Venus, and that it was in a retrograde orbit around the sun among the outer planets. He suggested that the orbit might not be 'exactly circular but somewhat oblong, like the figure commonly called oval.' This contradicted the Aristotelian doctrine that comets were meteorological phenomena occurring below the sphere of the moon. He was thus led to reject Aristotle's concept of the homocentric crystalline spheres, and he concluded that the planets were moving independently through space.

> There really are not any spheres in the heavens [...] Those which have been devised by the experts to save the appearances exist only in the imagination for the purpose of enabling the mind to conceive the motion which the heavenly bodies trace in their course and, by the aid

of geometry, to determine the motion numerically through the use of arithmetic.

Despite his admiration for Copernicus, Tycho rejected the heliocentric theory, both on physical grounds and on the absence of stellar parallax, where in the latter case he did not take into account the argument made by Archimedes and Copernicus that the stars were too far away to show any parallactic shift. Tycho rejected both the diurnal rotation of the earth as well as its annual orbital motion, retaining the Aristotelian belief that the stars rotated nightly around the celestial pole.

Faced with the growing debate between the Copernican and Ptolemaic theories, Tycho was led to propose his own planetary model, which combines elements of the geocentric and heliocentric theories. In the Tychonic system the immobile earth is still at the centre of the universe, with the sphere of the fixed stars revolving around it in 24 hours. Mercury and Venus orbit the sun, while Mars, Jupiter and Saturn orbit the stationary earth. (See the upper drawing in Figure 6.) Tycho believed that his model combined the best features of both the Ptolemaic and Copernican theories, since it kept the earth stationary and explained why Mercury and Venus were never very far from the sun.

Tycho's patron Frederick II died in 1588 and was succeeded by his son Christian IV, who was then eleven years old. When Christian came of age, in 1596, he informed Tycho that he would no longer support his astronomical research. Tycho was thus forced to abandon Uraniborg the following year, taking with him all of his astronomical instruments and records, hoping to find a new royal patron.

Tycho moved first to Copenhagen and then in turn to Rostock and Wandsburg Castle, outside Hamburg. He remained for two years at Wandsburg Castle, where in 1598 he published his *Astronomiae instauratae mechanica*, a description of all his astronomical instruments. He sent copies of his treatise to all of the wealthy and powerful people who might be interested in supporting his further researches; appending his star catalogue to the copy he presented to Emperor Rudolph II, who agreed to support Tycho's work, appointing him as the court astronomer.

Thus in 1600 Tycho moved to Prague, where he set up his instruments and created a new observatory at Benatky Castle, several miles

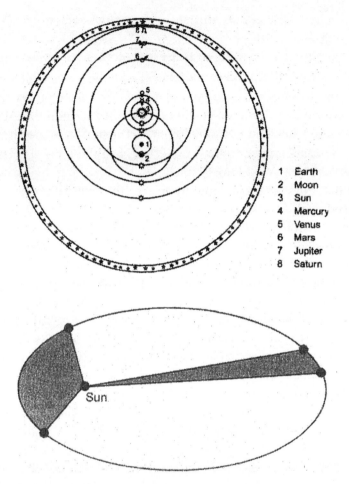

Figure 6 The Tychonic system (above); Kepler's first two laws of planetary motion (below).

northeast of the city. Soon afterwards he hired an assistant named Johannes Kepler (1571–1630), a young German mathematician who had sent him an interesting treatise on astronomy, the *Mysterium cosmographicum*.

Kepler was born on 27 December 1571 at Weil der Stadt in southwestern Germany. His father was an itinerant mercenary soldier, his mother a fortune-teller who at one point was accused of being a witch and almost burned at the stake. The family moved to the nearby town of Lemberg, where Kepler was enrolled in one of the excellent Latin schools founded by the Duke of Württemberg. His youthful interest in astronomy had been stimulated by seeing the comet of 1577 and a lunar eclipse in 1580.

In 1589 Kepler entered the University of Tübingen, where in addition to his studies in mathematics, physics and astronomy he was influenced by Platonism, Pythagoreanism and the cosmological ideas of Nicholas of Cusa. His mathematics lectures were based on the works of Euclid, Archimedes and Apollonius of Perge. (As Kepler later said, 'How many mathematicians are there, who would toil through the *Conics* of Apollonius of Perge?')

Kepler was particularly influenced by his professor of astronomy, Michael Maestlin, from whom he first learned of the heliocentric theory. Maestlin's popular textbook, *Epitome astronomiae*, does not mention the heliocentric theory. But in his lectures at the University of Tübingen he discussed the Copernican system, telling of how it accounted for retrograde motion and how it provided a model which explained the harmonious manner in which the earth and the other five planets orbited the sun, with periods that increased with their distance from the centre.

Kepler received his master's degree at Tübingen in 1591, after which he studied theology there until 1594, when he was appointed Teacher of Mathematics at the Protestant seminary in the Austrian town of Graz, a year after his arrival in the city.

Kepler developed an idea that he thought explained the arrangement and order of the heliocentric planetary system. He had learned from his reading of Euclid that there were five and only five regular polyhedra, the so-called Platonic solids, in which all of the faces are equal as well as equilateral – the cube, tetrahedron, dodecahedron, icosahedron and octahedron – and it occurred to him that they were related to the orbits of the earth and the five other planets. He explained the scheme in his treatise, the *Mysterium cosmographicum*, published in 1596:

> The earth's orbit is the measure of all things; circumscribe around it a dodecahedron, and the circle containing it will be Mars; circumscribe around Mars a tetrahedron, and the circle containing this will be Jupiter; circumscribe around Jupiter a cube, and the circle containing this will be Saturn. Now inscribe within the earth an icosahedron, and the circle contained in it will be Venus; inscribe within Venus an octahedron, and the circle contained in it will be Mercury. You now have the reason for the number of planets.

In the introduction to the *Mysterium cosmographicum,* Kepler wrote of his excitement on discovering the work of Copernicus, which he described as 'a still unexhausted treasure of truly divine insight into the magnificent order of the whole world and of all bodies.' He wrote also of how Copernicus inspired him to begin thinking about the causes behind this cosmic order:

> When I was studying under the distinguished Michael Maestlin at Tübingen six years ago, seeing the inconveniences of the commonly accepted theory of the universe, I became so delighted with Copernicus, whom Maestlin often mentioned in his lectures, that I often defended his opinions in the students' debates about physics. I even wrote a painstaking disputation about the first motion, maintaining that it happens because of the rotation of the earth. I have by degrees – partly out of hearing Maestlin, partly by myself – collected all the advantages that Copernicus has over Ptolemy. At last in the year 1595 in Graz when I had an intermission in my lectures, I pondered on the subject with the whole energy of my mind. And there were three things above all for which I sought the causes as to why it was this way and not another – the number, the dimensions, and the motion of the orbs.

The values that Kepler had for the relative radii of the planetary orbits agree fairly well with those determined by Copernicus, when allowances are made for the eccentricities of the orbits. Also, in the case of Mercury, he compromised by taking the radius of a sphere formed by the edges of the octahedron rather than in the octahedron itself though there was no physical basis for his theory. With these concessions everything agrees within five per cent, except for Jupiter, where the Kepler's value is off by nine per cent, at which, as he writes, 'no one will wonder, considering such a great distance.'

Kepler sent copies of his treatise to a number of scientists, including Galileo Galilei (1564–1642). In his letter of acknowledgement, dated 4 August 1597, Galileo congratulated Kepler for having had the courage, which he himself lacked, of publishing a work supporting the Copernican theory.

Kepler wrote back to Galileo on 13 October 1597, encouraging him to continue supporting the Copernican theory. 'Have faith, Galilei, and come forward!' he wrote. 'If my guess is right, there are but few of the prominent mathematicians of Europe who would wish to secede from us: such is the power of truth.'

Kepler had also sent a copy of the *Mysterium Cosmographicum* to Tycho Brahe, who received it after he had left Denmark for Germany. Tycho responded warmly, calling the treatise 'a brilliant speculation,' beginning a correspondence that eventually led Kepler to accept his invitation to join him at his new observatory at Benatky. As Tycho wrote in response to Kepler's letter of acceptance: 'You will come not so much as a guest but as a very welcome friend and highly desirable participant and companion in our observations of the heavens.'

Kepler finally arrived at Benatky with his family in late December 1599, beginning a brief but extraordinarily fruitful collaboration with Tycho. When Kepler began work at Benatky he had hopes that he could take Tycho's data and use it directly to check his own planetary theory. But he was disappointed to find that most of Tycho's data was still in the form of raw observations, which first had to be subjected to mathematical analysis. Moreover Tycho was extremely possessive of his data and would not reveal any more of it than Kepler needed for his work.

These and other disagreements with Tycho led Kepler to leave Benatky in April of that year, though he returned in October after considerable negotiation concerning the terms of his employment. Tycho then assigned Kepler the task of analysing the orbit of Mars, which up to that time had been the responsibility of his assistant Longomontanus, who had just resigned. As Kepler wrote in his *Astronomia nova*, published in 1609:

> Tycho Brahe, himself an important part in my destiny, continually urged me to come to visit him. But since the distance of the two places would have deterred me, I ascribe it to Divine Providence that he came to Bohemia. I arrived there just before the beginning of the year 1600 with the hope of obtaining the correct eccentricities of the planetary orbits. Now at that time Longomontanus had taken up the theory of Mars, which was placed in his hands so that he might study the Martian opposition with the Sun in 9° of Leo. Had he been occupied with another planet, I would have started with that same one. That is why I again consider it an effect of Divine Providence that I arrived in Prague at the time when he was studying Mars, because for us to arrive at the secret knowledge of astronomy, it is absolutely necessary to use the motion of Mars, otherwise that knowledge would remain eternally hidden.

Mars and Mercury are the only visible planets with eccentricities large enough to make their orbits significantly different from perfect circles. But Mercury is so close to the sun that it is difficult to observe, leaving Mars as the ideal planet for checking a mathematical theory, which is why Kepler was so enthusiastic at being able to analyse its orbit.

Early in the autumn of 1601 Tycho brought Kepler to the imperial court and introduced him to Emperor Rudolph. Tycho then proposed to the emperor that he and Kepler compile a new set of astronomical tables. With the emperor's permission, this would be named the *Rudolphine Tables*, and since it was to be based on Tycho's observations it would be more accurate than any done in the past. The emperor graciously consented and agreed to pay Kepler's salary in this endeavour.

Soon afterwards Tycho fell ill, and after suffering in agony for eleven days he died on 24 October 1601. On his deathbed he made Kepler promise that the *Rudolphine Tables* would be completed, and he expressed his hopes that it would be based on the Tychonic planetary model. As Kepler later wrote of Tycho's final conversation with him: 'although he knew I was of the Copernican persuasion, he asked me to present all my demonstrations in conformity with his hypothesis.'

Two days after Tycho's death the Emperor Rudolph appointed Kepler as Court Mathematician and Head of the Observatory at Benatky. Kepler thereupon resumed his work on Mars, now with unrestricted access to all of Tycho's data. At first he tried the traditional Ptolemaic methods – epicycle, eccentric and equant – but no matter how he varied the parameters the calculated positions of the planet disagreed with Tycho's observations by up to eight minutes of arc. His faith in the accuracy of Tycho's data led him to conclude that the Ptolemaic theory of epicycles, which had been used by Copernicus, would have to be replaced by a completely new theory, as he wrote: 'Divine Providence granted us such a diligent observer in Tycho Brahe, that his observations convicted this Ptolemaic calculation of an error of eight minutes; it is only right that we should accept God's gift with a grateful mind [...] Because those eight minutes could not be ignored, they alone have led to a total reformation of astronomy.'

After eight years of intense effort Kepler was finally led to what is now known as the second of his two laws of planetary motion. The second law states that a radius vector drawn from the sun to a planet sweeps out equal areas in equal times, so that when the planet is close to the sun it moves rapidly and when far away it goes slowly. This law is first correctly stated in book V of his *Epitome astronomiae Copernicanae*, published in 1621. The law worked well for the earth's orbit, but when it was applied to Mars the eight-minute discrepancy once again appeared, which made Kepler realize that the planetary orbits might not be circular.

Kepler knew that the epicycle theory for Mercury gave an ovoid curve, but when he tried this for Mars the discrepancy was still four minutes of arc. Seeing that the ovoid curve he had drawn for the orbit of Mars was quite similar to an ellipse, he began thinking of the possibility that the planetary orbits were elliptical. As he wrote to his friend David Fabricius in July 1603: 'I lack only a knowledge of the geometric generation of the oval or face-shaped curve [...] If the figure were a perfect ellipse, then Archimedes and Apollonius would be enough.' He soon realized that an ellipse would satisfy the calculations, but such an orbit did not fit in with his idea that the motion of the planets was driven by the magnetic field of the rotating sun. He expresses his frustration in chapter 58 of his *Astronomiae Nova*, the book that he had begun preparing in 1604:

> I was almost driven to madness in considering and calculating this matter. I could not find out why the planet would rather go on an elliptical orbit. Oh, ridiculous me! As the libration in the diameter could not also be the way to the ellipse. So this notion brought me up short, that the ellipse exists because of the libration. With reasoning derived from physical principles, agreeing with experience, there is no figure left for the orbit of the planet except a perfect ellipse.

Thus he arrived at what is now known as Kepler's first law of planetary motion, which is that each of the planets travels in an elliptical orbit, with the sun at one of the two focal points of the ellipse. The second law states that a radius vector drawn from the sun to a planet sweeps out equal areas in equal times, so that when the planet is close to the sun it moves rapidly and when farther away it goes slowly. (See the lower drawing in Figure 6.) These two laws, which first appeared in Kepler's *Astronomia nova*, became the basis for his subsequent work on the *Rudolphine Tables*.

Kepler's first two laws of planetary motion eliminated the need for the epicycles, eccentrics, and deferents that had been used by astronomers from Ptolemy to Copernicus. The passing of this ancient cosmological doctrine was noted by Milton in Book VIII of *Paradise Lost*.

> Hereafter, when they come to model Heaven,
> And calculate the stars; how they will wield
> The mighty frame; how build, unbuild, contrive
> To save appearances; how gird the sphere
> With centric and eccentric scribbled o'er,
> Cycle and epicycle, orb in orb.

Kepler wrote three other works on his researches before the publication of his *Astronomia nova*. The first was *Astronomomiae pars optica*, published in 1603, and the second was *Ad Vitellionem paralipomena* (Appendix to Witelo), which came out the following year. Both books dealt with optical phenomena in astronomy, particularly parallax and refraction, as well as the annual variation in the size of the sun.

The third book was occasioned by another new star that appeared in October 1604 in the vicinity of Jupiter, Saturn and Mars. Kepler published an eight-page tract on the new star in 1606 entitled *De stella nova*, with a subtitle describing it as 'a book full of astronomical, physical, metaphysical, meteorological, astrological discussions, glorious and unusual.' At the end of the tract Kepler speculated on the astrological significance of the new star, saying that it might be a portent of the conversion of the American Indians, a mass migration to the New World, the downfall of Islam, or even the second coming of Christ.

Meanwhile the whole science of astronomy had been profoundly changed by the invention of the telescope. Instruments called perspective glasses had been used in England before 1580 for viewing distant terrestrial objects, and both John Dee and Thomas Digges were known to be expert in their construction and use, though there is no evidence that they used them for astronomical observations. But their friend Thomas Harriot is known to have made astronomical observations in the winter of 1609–10 with a small telescope, which may have been a perspective glass.

Other than these perspective glasses, the earliest telescope seems to have appeared in 1604, when a Dutch optician named Zacharias Janssen constructed one from a specimen belonging to an unknown Italian, after which he sold some of them at fairs in northern Europe. When Galileo heard of the telescope, he constructed one in his workshop in 1609, after which he offered it to the Doge of Venice for use in war and navigation. After improving on his original design, he began using his telescope to observe the heavens, and in March 1610 he published his discoveries in a little book called *Siderius nuncius* (The Starry Messenger).

Galileo sent a copy of the *Siderius nuncius* to Kepler, who received it on 8 April 1610. During the next 11 days Kepler composed his response in a little work called *Dissertatio cum Nuncio sidereal* (Answer to the Sidereal Messenger), in which he expressed his enthusiastic approval of Galileo's discoveries and reminded readers of his own work on optical astronomy, as well as speculating on the possibility of inhabitants on the moon and arguing against an infinite universe. Galileo wrote a letter of appreciation to Kepler, saying that 'I thank you because you are the first one, and practically the only one, to have complete faith in my assertions.'

Kepler borrowed a telescope from the Elector Ernest of Cologne at the end of August 1610, and for the next 10 days he used it to observe the heavens, particularly Jupiter and its moons. He published the results the following year in a booklet entitled *Narratio de Jovis satellitibus*, confirming the authenticity of Galileo's discovery.

Kepler's excitement over the possibilities of the telescope was such that he spent the late summer and early autumn of 1610 making an exhaustive study of the passage of light through lenses, which he published later that year under the title *Dioptrice*, which became one of the foundation stones of the new science of optics.

The death of Rudolph II early 1612 forced Kepler to leave Prague and take up the post of district mathematician at Linz, where he remained for the next 14 years. During the period that Kepler lived in Linz he continued his calculations on the *Rudolphine Tables* and published two other major works, the first of which was *Harmonice mundi* (Harmony of the World), which appeared in 1619. The title was inspired by a Greek manuscript of Ptolemy's treatise on musical theory, the *Harmonica*, which Kepler acquired in 1607 and used in his analysis of music, geometry, astronomy and astrology. The most important part of the *Harmonice* is the relationship now known as Kepler's Third Law of Planetary Motion, which he discovered on 15 May 1618, and presents in Book V. The law states that for each of the planets the square of the period of its orbital motion is proportional to the cube of its mean distance from the sun.

There had been speculations about the relation between the periods of planetary orbits and their radii since the time of Pythagoras, and Kepler was terribly excited that he had at last, following in the footsteps of Ptolemy, found the mathematical law 'necessary for the contemplation of celestial harmonies.' He wrote of his pleasure 'that the same thought about the harmonic formulation had turned up in the minds of two men (though lying so far apart in time) who had devoted themselves entirely to contemplating nature [...]I feel carried away and possessed by an unutterable rapture over the divine spectacle of the heavenly harmony.'

Kepler dedicated the *Harmonice* to James I of England. The King responded by sending his ambassador Sir Henry Wooton with an invitation for Kepler to take up residence in England. But after considering the offer for a while Kepler eventually decided against it.

The English poet John Donne was familiar with the work of Copernicus and Kepler, probably through Thomas Harriot. Donne had in 1611 said to the Copernicans that 'those opinions of yours may very well be true [...]creeping into every man's mind.' That same year Donne lamented the passing of the old cosmology in 'The Anatomy of the World':

> And new Philosophy calls all in doubt,
> The Element of fire is quite put out;
> The Sun is lost, and th' earth, and no man's wit
> Can well direct him, where to look for it.

Kepler's second major work at Linz was his *Epitome astronomiae Copernicanae* (Epitome of Copernican Astronomy) in seven books, the first three of which were published in 1617, the fourth in 1620, and the last three in 1621. In the first three of the seven books of the *Epitome* Kepler refutes the traditional arguments against the motions of the earth, going much farther than Copernicus and using principles that Galileo would later give in greater detail.

Kepler's three laws of planetary motion are explained in great detail in Book IV, along with his lunar theory. The last three books treat practical problems involving his first two laws of planetary motion as well as his theories of lunar and solar motion and the precession of the equinoxes. The work became very popular, and for the next two decades the *Epitome* was the most widely read treatise on theoretical astronomy in Europe.

Kepler was forced to leave Linz in 1626 and move to Ulm, from where he moved on to Sagan in 1628. While Kepler was still in Ulm he published the *Rudolphine Tables* in September 1627, dedicating them to Archduke Ferdinand II. In the preface Kepler explained that the long delay in the publication of the tables, noting that 'the novelty of my discoveries and the unexpected transfer of the whole of astronomy from fictitious circles to natural causes were most profound to investigate, difficult to explain, and difficult to calculate, since mine was the first attempt.'

A decade earlier Kepler had come across the pioneering work on logarithms published in 1614 by the Scottish mathematician John Napier (1550–1617). He used logarithms in computing the planetary positions in the *Rudolphine Tables*, appending logarithmic tables as well as a catalogue of the celestial coordinates of a thousand stars, all based on the Copernican theory. The new tables were far more accurate than any in the past, and they remained in use for more than a century.

Kepler departed from Sagan on 8 October 1630 to go to a book fair in Leipzig, after which he went on to attend a meeting in Regensburg, where he arrived on 2 November. He became ill on the journey and died in Regensberg of an acute fever on 15 November 1630. His tombstone, now lost, was engraved with an epitaph that he had written himself: 'I used to measure the heavens, now I measure the shadow of the earth./Although my soul was from heaven, the shadow of my body lies here.'

Kepler used his tables to predict that Mercury and Venus would make transits across the disk of the sun in 1631. The transit of Venus was not observed in Europe because it took place at night. The transit of Mercury was observed by Pierre Gassendi in Paris on 7 November 1631, representing a triumph for Kepler's astronomy, for his prediction was in error by only 10 minutes of arc as compared to 5 degrees for tables based on Ptolemy's model.

Kepler's last work, written between 1620 and 1630 and published posthumously in 1634, was a curious tract called *Somnium, seu opus posthumum de astronomia lunari* (Dream on Lunar Astronomy), which has been described as the first work of science fiction. It tells of how a student of Tycho Brahe is transported by occult forces to the moon, from where he describes the earth and its motion around the sun along with the other planets. Kepler's notes for the *Somnium* attributes the tides 'to the bodies of the sun and moon attracting the waters of the sea with a certain force similar to the magnetic.'

Kepler's three laws of planetary motion were to become the basis of the new heliocentric astronomy that emerged in the seventeenth century, culminating in Newton's *Principia*, published in 1687. When the *Principia* was first introduced to the Royal Society in London it was described as 'a mathematical demonstration of the Copernican hypothesis as proposed by Kepler.' The astronomer Edmond Halley, in his review, wrote that Newton's first eleven propositions in Book I of the *Principia* were 'found to agree with the Phenomena of the Celestial Motions, as discovered by the great Sagacity and Diligence of Kepler.'

CHAPTER 14
DEBATING THE COPERNICAN AND PTOLEMAIC MODELS

Despite the revolutionary ideas and discoveries of Copernicus, Tycho Brahe, Kepler and others, the Aristotelian geocentric model of the universe and the planetary theory of Ptolemy still formed the general world-picture in the early seventeenth century. But by the end of the century the heliocentric model of Copernicus had gained general acceptance and the geocentric model of Aristotle and Ptolemy had been discredited, principally through the efforts of Galileo Galilei (1564–1642), whose works opened the way for the new physics and astronomy of Newton.

Galileo was born in Pisa on 15 February 1564, the first son of Vincenzio Galilei of Florence and Giula Ammananti of Pescia, who moved back to Florence in 1574. Vincenzio was a musician and musical theorist, who in 1589 published a work on the numerical theory of musical harmony, based on an experimental study of consonance and its relation to the lengths and tensions of strings. Galileo's interest in the testing of laws of mathematical physics by observation probably stems from his father's musical investigations.

Galileo received his elementary education from a private tutor, Jacopo Borghini, before being sent to school at the renowned Camaldolese monastery of Santa Maria at Vallombrosa. But when he joined the order as a novice his father brought him back to Florence, where he studied at a school run by the Camaldolesa monks, though not as a novice in their order.

Galileo was enrolled in the school of medicine at the University of Pisa in 1581, studying physics and astronomy under Francesco Buonamici, philosophy with Girolamo Borro, and mathematics, including astronomy, with Filippo Fantoni. He left Pisa without a degree in 1585 and returned to Florence, where he began an independent study of Euclid and Archimedes under Ostilio Ricci, the Tuscan court mathematician.

During the years 1585–9 Galileo gave private lessons in mathematics at Florence as well as private and public instruction at Siena. He travelled to Rome in 1587 to meet Christopher Clavius, the renowned Jesuit mathematician at the Collegio Romano, who for the rest of his days would be a friend and supporter of Galileo.

Galileo had in 1583 made his first scientific discovery, that the period of a pendulum is independent of the angle through which it swings, at least for small angles. He is supposed to have come to this conclusion by observing the oscillation of a chandelier in a cathedral. Three years later he invented an hydraulic balance, which he described in his first scientific publication, *La Balancetta* (The Little Balance), based on Archimedes' principle, which he also used in determining the centre of gravity of solid bodies.

Galileo was appointed Professor of Mathematics in 1589 at the University of Pisa, where he remained for only three years. During this period he wrote an untitled treatise on motion now referred to as *De Motu* (On Motion), which remained unpublished during his lifetime. The treatise was an attack on Aristotelian physics, such as the notion that heavy bodies fall more rapidly than light ones, which Galileo is supposed to have refuted by dropping weights from the Leaning Tower of Pisa, though there is reason to believe that this was merely a 'thought experiment.'

A similar experiment had already been performed in 1586 by the Flemish engineer Simon Stevin (1548–1620), court mathematician to Prince Maurice of Orange. Stevin took two lead balls, one ten times the weight of the other and dropped them thirty feet onto a wooden board from the church tower in Delft, concluding from the sound that they reached the ground at the same time, thus refuting Aristotle's theory.

Stevin was the first to establish the basic laws of hydrostatics and statics, in both cases beginning with the assumption that perpetual motion is impossible. He showed that the pressure of a liquid on the base of the containing vessel depended only on depth and was independent of shape and volume, the basic law of hydrostatics.

Stevin's writings include a pioneering work on musical theory, published in 1585. This was apparently inspired by the writings of Vincenzio Galilei, Galileo's father. Galileo in turn would be inspired by Stevin's writings, a remarkable cycle which says much about the developing character of science in the late sixteenth century.

Galileo's *De Motu* was probably written in opposition to a massive work of the same title by his teacher Francesco Buonamici, who had first introduced him to the study of motion. Although Galileo rejected many of Buonamici's ideas, which were Aristotelian, he was deeply influenced by his teacher and was in substantial agreement with him in the general methodology they used in their theory of motion, which they put on an axiomatic basis as in mathematics.

While teaching at Pisa Galileo was influenced by the writings of Giovanni Battista Benedetti, who was highly critical of Aristotle's ideas on motion. One of the Aristotelian concepts he rejected was that heavier bodies fall faster than lighter ones. Galileo, in his *De Motu*, rejected the Aristotelian theory in which bodies fall or rise to their natural place, with earth at the centre and above it in succession water, air and fire. Instead, he said that whether a body moves downward or upward depends on its density relative to that of the medium in which it is moving. He then argued that these phenomena should be seen as balance problems, so that the ninth section in *De Motu* is entitled 'In which all that was demonstrated above is considered in physical terms, and bodies moving naturally are reduced to weights on a balance,' a method he credited to Archimedes.

Thus Galileo proposed a method for solving all problems of motion, including the motions of floating bodies that could be reduced to that of weights on an Archimedian balance. He went on to show that all simple machines, such as the lever, the inclined plane and the pendulum, could also be reduced to balance problems. He interpreted the problem of free fall as an instance of floating bodies or a balance that had no weight on one side.

In 1592 Galileo was appointed to the chair of mathematics at the University of Padua, where he remained for 18 years. During his Paduan residency Galileo wrote several treatises for the use of his students, including one that was first published in a French translation in 1634 under the title *Le meccaniche,* a study of motion and equilibrium on inclined planes that further developed the ideas he had presented in *De Motu.* Here he bridged the gap between statics and dynamics by remarking that an infinitesimal force would be sufficient to disturb equilibrium.

In *Le meccaniche* Galileo used the concept of centre of gravity to discuss simple machines, as when he says 'Thus, *moment* [*momento*] is the impetus to go downward composed of heaviness, position and of anything else by which this tendency may be caused.' The model that he employed is a single-arm balance or lever, where moment is a generalized force. In the case of a balance, moment is the product of weight times the perpendicular distance between the line of action of the weight and the fulcrum, while for a body moving on an inclined plane the moment depends on the angle of inclination of the plane. Using this model, he went on to explain the lever and other types of machines before finally making a first attempt to deal with the force of impact, all in the Archimedean tradition.

In May 1597 Galileo wrote to a former colleague at Pisa defending the Copernican theory. Three months later he received a copy of *Mysterium cosmographicum,* which led to his first correspondence with Kepler. In his letter of acknowledgement to Kepler on 4 August 1597, Galileo wrote:

> Many years ago I came to agree with Copernicus, and from this position the causes of many natural effects have been found by me which doubtless cannot be explained by the ordinary supposition. I wrote down many reasons and arguments, which, however, I did not venture until now to divulge, deterred by the fare of Copernicus himself, our master, who, although having won immortal fame with some few, to countless others appears [...] as an object of derision and contumely. Truly, I would venture to publish my views if more like you existed; since this is not so, I will abstain.

Around the same time Galileo constructed a mathematical instrument which he called the 'geometric and military compass,' later called the 'sector,' which served among other purposes that of the instrument now known as the proportional divider. He hired a skilled artisan to make this and other instruments in his own workshop for sale. In 1606 he published his first work, a handbook of instructions for users of the compass, dedicating it to the young prince Cosimo II de' Medici, to whom he had been giving private lessons.

In 1604 Galileo wrote a letter to his friend Paolo Sarpi, saying that on the basis of an axiom he had proved that the distances covered by a falling body are proportional to the square of the times. This is the law of kinematics now written as $s = \frac{1}{2}at^2$, where s is the distance, a is the acceleration, and t is the time. The axiom that he adopted was that, for a body falling from rest, the instantaneous velocity, that is, the velocity at any moment in time is proportional to the distance traversed, which is not true. The fact that the acceleration was proportional to the time rather than the distance had first been stated unequivocally by the Spanish Dominican friar Domingo de Soto (1494–1560). Thus Galileo had derived the correct law giving the distance covered as a function of time for a body moving under the influence of gravity, though he based his derivation on an erroneous assumption. Galileo later corrected this mistake, apparently after carrying out an experiment to test the law with a bronze ball rolling down an inclined plane. He defined equal intervals of time as those during which equal weights of water issued from a small hole in a bucket; he used a very large amount of water relative to the amount issuing through the hole, so that the decrease in head was unimportant.

The year 1609 was a turning point in Galileo's career, for it was then that he first learned of the existence of the telescope. He quickly made one for himself, eventually developing an improved model with a magnifying power of thirty. The following year he described his observations in the *Siderius nuncius* (The Starry Messenger), dedicated to Cosimo II de' Medici, who by then had succeeded as Grand Duke of Tuscany.

After the dedication Galileo gives a brief summary of the book, which in its English translation reads:

THE STARRY MESSENGER: Revealing great, unusual, and remarkable spectacles, opening these to the consideration of every man, and especially of philosophers and astronomers; as Observed by Galileo Galilei, Gentleman of Florence, Professor of Mathematics at the University of Padua, With the Aid of a Spyglass lately invented by him, in the surface of the Moon, in innumerable Fixed Stars, in Nebulae, and above all, in FOUR

PLANETS swiftly revolving about Jupiter at differing distances and periods, and known to no one before the Author recently perceived them and decided that they should be named THE MEDICIAN STARS.

The 'Medicean Stars' were the four principal moons of Jupiter that Galileo discovered with his telescope. He called them 'planets' because they circled Jupiter just as it and the other planets orbit the sun, as he writes in the dedication, the first public notice by Galileo that he accepted the Copernican system:

> Behold, then, four stars to bear your famous name; bodies which belong not to the incongruous multitude of fixed stars, but to the bright ranks of the planets. Variously moving about most noble Jupiter as children of his own, they complete their orbits with marvelous velocity – at the same time executing with one harmonious accord mighty revolutions every dozen years about the center of the universe, that is, the sun.

Galileo begins the book with an account of his systematic observation of the moon, which he which he describes in great detail and illustrates with drawings and geometrical diagrams. As he writes of his first lunar observation: 'On the fourth or fifth day after new moon, when the moon is seen with brilliant horns, the boundary which divides the dark part from the light does not extend uniformly in an oval line as would happen on a perfectly spherical solid, but traces out an uneven, rough and very wavy line, as shown in the figure below.' He concluded that 'the surface of the moon is not smooth, and precisely spherical as a great number of philosophers believe it (and the other heavenly bodies) to be, but is uneven, rough, and full of cavities and prominences, being not unlike the face of the earth, relieved by chains of mountains and deep valleys.'

He then writes of his survey of the planets and stars, noting that with his telescope the planets were magnified and appeared as illuminated globes, looking like small moons, while the fixed stars still appeared as brilliant points of scintillating light, though much brighter than when viewed with the naked eye:

> Deserving of notice also is the difference between the appearance of the planets and of the fixed stars. The planets show their globes perfectly round and definitely bounded, looking like little moons, spherical and

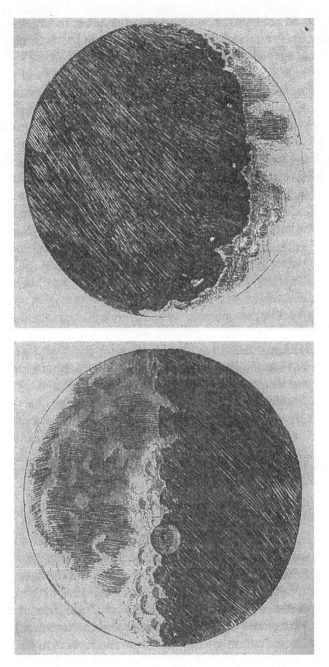

Figure 7 Galileo's observations of the Moon with the telescope, from *Siderius nuncius* (The Starry Messenger), 1610.

flooded all over with light; the fixed stars are never seen to be bounded by a circular periphery, but have rather the aspect of blazes whose rays vibrate about them and scintillate a great deal. Viewed with a telescope they appear of a shape similar to that which they present to the naked eye, but sufficiently enlarged so that a star of the fifth or sixth magnitude seems to equal the Dog Star, the largest [actually not the largest but the brightest] of all the fixed stars.

He saw that in addition to stars of the sixth magnitude, the faintest of those that can be seen with the naked eye, 'a host of other stars are perceived through the telescope which escape the naked eye; these are so numerous as almost to surpass belief.

He observed the constellation Orion, intending to depict the new stars he had observed along with the familiar ones visible to the naked eye, but he says that he 'was overwhelmed by the vast quantity of stars and by the limitations of time, so I have deferred this to another occasion. There are more than 500 new stars distributed among the old ones within limits of one or two degrees of arc.' Thus, as he writes, in depicting Orion in his book, he limited the scope of his drawing to the sword and belt of Orion, showing 80 new stars in addition to the nine visible to the naked eye:

> Hence to the three stars in the Belt of Orion and the six in the Sword which were previously known, I have added eighty adjacent stars discovered recently, preserving the intervals between them as exactly as I could. To distinguish the known or ancient stars, I have depicted them larger and outlined them doubly; the other (invisible) stars I have drawn smaller and without the extra line. I have also preserved differences of magnitude as well as possible.

He did the same with the Pleiades, where in addition to the six brightest of the seven visible stars of the constellation he depicted 36 new ones in the illustration. He then turned his attention to the Milky Way, where, as he notes, he resolved questions about its nature and composition that had been argued since antiquity:

> With the aid of the telescope this has been scrutinized so directly and with such ocular certainty that all the disputes which have vexed philosophers through so many ages have been resolved, and we are at last freed from wordy disputes about it. The galaxy is, in fact, nothing but

> a congeries of innumerable stars grouped together in clusters. Upon whatever part of it the telescope is turned a vast crowd of stars is immediately presented to view. Many of them are rather large and quite bright, while the number of smaller ones is quite beyond calculation.

Besides the Milky Way, he observed other celestial 'clouds' that in the past had been 'nebulous,' and he found that they too turned out to be clusters of previously invisible stars. He depicted two of these star clusters in his illustrations, the Nebula of Orion and the Nebula of Praesepe, which lies in the constellation Cancer.

Galileo made an improved version of his telescope with a higher magnification, and with this, on 7 January 1610, he discovered what at first he thought to be small stars close to Jupiter and aligned with it, two on the eastern side and one on the west. The following night he found that the three 'starlets', as he called them, were all to the west of Jupiter and aligned at equal intervals from one another and the plane. Observations on subsequent nights revealed the starlets changing their orientations with respect to one another and to Jupiter, though still forming a straight line. On 10 January only two of them appeared, both of them to the east of Jupiter, leading Galileo to suppose that the third was hidden behind the planet. On 13 January he saw for the first time four starlets, one easterly and the others easterly, the middle western star slightly to the north of the others. He eventually concluded that the four starlets were in fact satellites of Jupiter, those closest to the planet revolving around it more swiftly than those farther away, just as the planets orbit the sun, which he presented as a conclusive argument in favour of the Copernican system:

> Here we have a fine and elegant argument for quieting the doubts of those who, while accepting with quiet tranquility the revolutions of the planets about the sun in the Copernican system, are mightily disturbed to have the moon alone revolve around the earth and accompany it in an annular rotation around the sun. Some have believed that this structure of the universe should be rejected as impossible. But now we have not just one planet rotating about another while both run through a great orbit around the sun; our own eyes show us four stars which wander around Jupiter as does the moon around the earth, while all together trace out a grand revolution about the sun in the space of 12 years.

Cosimo II de' Medici responded by making Galileo his court philosopher and appointing him to the chair of mathematics at the University of Pisa. Galileo had no obligation to teach at the University of Pisa or even to reside in the city, and so after his appointment, in September 1610, he departed to take up residence in Florence.

Meanwhile Galileo had made two more important discoveries with his telescope. The first was the curious shape of Saturn, which appeared to have a pair of equatorial protuberances. These were actually the famous rings of Saturn, which his primitive telescope was unable to resolve, and he thought that they were a pair of small satellites in stationary orbits close to the planet. The second discovery was that Venus goes through a cycle of phases similar to those of the moon, the only difference being that its size varied as its distance from the earth changed. Galileo showed that this was due to the fact that Venus was in orbit around the sun rather than the earth. He communicated these discoveries to Kepler and to the Jesuit astronomers in Rome as added evidence in support of the Copernican theory.

Late in 1610 Galileo received a congratulatory letter from his friend Father Christopher Clavius at the Collegio Romano in Rome, informing him that the astronomers there had verified his discovery of the new fixed stars and the satellites of Jupiter. His colleague, Father Grienberger, wrote in January 1621 to Galileo: 'Things so hard to believe as what you assert neither can nor should be believed lightly; I know how difficult it is to dismiss opinion sustained for many centuries by the authority of so many scholars. And surely, if I had not seen, so far as the instruments allowed, these wonders with my own eyes [...] I do not know whether I would have consented to your arguments.'

Thus encouraged, in March 1611 Galileo visited Rome, where he was welcomed by noblemen and by church dignitaries and was given a friendly interview by Pope Paul V. During Galileo's stay in Rome he was elected to the Accademia dei Lincei (Academy of the Lynx-Eyed), founded in 1603 by Federigo Cesi, Marquis of Montecelli.

Cardinal Robert Bellarmine, head of the Collegio Romano, asked the mathematicians on the faculty to evaluate Galileo's discoveries. When they confirmed them, Clavius and his colleagues honoured Galileo with a full day of ceremonies at the college, during which one of them delivered an oration praising him and his book.

But, as Stillman Drake writes, 'about the time Galileo left Rome to return to Florence, a letter went secretly to the chief inquisitor at Padua upon instructions from Bellarmine and six of his cardinals. It contained the ominous words: "see if Galileo is mentioned in the proceedings against Dr Cesare Cremonino."' Cremonino, head of the philosophy department at the University of Padua, was a good friend of Galileo, despite their philosophical differences. He was long suspected of heretical opinions by the inquisitors, but he was never brought to trial because he remained in Venetian territory.

Despite the confirmation of the Jesuit astronomers in Rome, conservative Aristotelians at the universities remained sceptical of Galileo's discoveries. Giulio Libri, who taught philosophy both at Pisa and Padua while Galileo was at those universities, refused even to look through a telescope. When Libri died in 1610, Galileo expressed the hope that since he had refused to look at the celestial bodies through a telescope when he was on earth, he could now see them with his own eyes on the way to heaven.

Galileo had resumed his earlier researches on hydrostatics in 1609, and three years later he published, in Italian, a treatise entitled *Discourse on Floating Bodies*. Here, as Stillman Drake notes, 'Using the concept of moment and the principle of virtual velocities, Galileo extended the scope of the Archimedean work beyond purely hydrostatic considerations.' This was a continuation and extension of his earlier researches in hydrostatics, to which he had returned partly because of a dispute concerning floating bodies in which he was involved with Lodovico delle Colombe, leader of a group of Aristotelians who were highly critical of his efforts in support of the Copernican theory.

Meanwhile Galileo had become embroiled in another dispute. This involved a work by a German astronomer, Father Christopher Scheiner, a Jesuit professor at the University of Ingoldstadt, who had observed spots on the sun. Scheiner's superior had forbidden him to publish his findings under his own name, and so he presented them in the form of several letters addressed to his friend Mark Welser, a wealthy merchant of Augsburg. Welser sent the letters to Galileo, with whom he had corresponded earlier concerning the lunar mountains, asking his opinion, concealing Scheiner's name under the pseudonym 'Apelles'. Galileo responded with a series of three letters, written in Italian since that was the language in which Welser had addressed him.

Scheiner had the first of Galileo's letters translated into German, after which he wrote a reply entitled *A More Accurate Discussion of Sunspots and the Stars which Move around Jupiter*, to which Galileo replied in his third letter to Welser.

Scheiner's first recorded observation of sunspots was made on 21 October 1611. Galileo's earliest known mention of the phenomenon occurs in a letter dated October 1621, in which he indicates that he had been observed sunspots 18 months earlier and had shown them to others while in Rome. He was apparently unaware that the Englishman Thomas Harriot had already reported his observation of sunspots, as had the German Johannes Fabricius, who announced his discovery in a booklet printed in the summer of 1611.

In all three of his letters Galileo gave detailed evidence countering Scheiner's theory that the spots were actually tiny planets orbiting close to the sun. Then, after giving an extremely detailed description of the phenomenon, he presented his own, correct, view that sunspots were part of the fluid solar surface and were in rotation about the sun's axis along with the rest of the solar sphere.

In the first letter Galileo also gave a detailed description of the moons of Jupiter as well his two more recent discoveries, that is, the phases of Venus and the protuberances on Saturn's disk, which he thought to be small moons. Then in his third letter he presented these discoveries as additional evidence for the Copernican theorem, predicting that it would soon be universally adopted. As he wrote, referring to the protuberances on Saturn: 'I say, then, that I believe that after the winter solstice of 1614 they may once more be observed. And perhaps this planet also, no less than horned Venus, harmonizes admirably with the great Copernican system, to the universal revelation of which doctrine propitious breezes are now seen to be directed towards us, leaving little fear of clouds or crosswinds.'

The second letter also contains Galileo's first published mention of the concept of inertia, according to which a body will preserve a state of uniform linear motion or of rest unless acted upon by a force. He states this principle in discussing the rotation of sunspots on the solar surface: 'For I seem to have observed that physical bodies have physical inclination to some motion (as heavy bodies downward), which motion is exercised by them through an intrinsic property and without need of a particular external mover, whenever they are not impeded by some obstacle.'

Galileo's *Letters on Sunspots*, containing his entire correspondence on the subject with Welser, was printed at Rome in 1613 under the auspices of the Lincean Academy. Since it was in Italian, as Stillman Drake remarks, it 'thus brought the question of the earth's motion to the attention of virtually everyone in Italy who could read.'

Meanwhile Galileo had been active in advancing the cause of Copernicanism against the accepted cosmology of Aristotle, which in its reinterpretation by St Thomas Aquinas formed part of the philosophical basis for Roman Catholic theology. This led to severe criticism of Galileo by his enemies, who interpreted his arguments as being attacks against the Church itself. On 16 December 1611 Galileo's friend Lodovico Cigoli wrote to him from Rome about a conspiracy to denounce him publicly:

> I have been told by a friend of mine, a priest who is very fond of you, that a certain crowd of ill-disposed men envious of your virtue and merits meet at the house of the archbishop there and put their heads together in a mad quest for any means by which they could damage you, either with regard to the motion of the earth or otherwise. One of them wished to have a preacher state from the pulpit that you are asserting outlandish things.
>
> The priest having perceived the animosity against you, replied as a good Christian and a religious man ought to do. Now I write this to you so that your eyes will be open to such envy and malice on the part of evildoers.

The climax of these attacks came on 21 December 1614, when the Dominican priest Thomas Caccini denounced Galileo, the Copernican system, and mathematics in general as being contrary to Christianity. Galileo's response came in the form of a letter to the Grand Duchess Christina, mother of Cosimo II de' Medici, who had expressed interest in his views. This work was probably completed by June 1615, though it was not published until 1636, under the title *Letter to Madame Christina of Lorraine, Grand Duchess of Tuscany, Concerning the use of Biblical Quotations in Matters of Science.* Here he argued that sacred scripture had to be interpreted in terms of what science knew about the world. Neither the Bible nor nature could speak falsely, he said, but the investigation of nature was the duty of scientists, and theologians should reconcile scientific facts with the language of the scriptures, noting that the Bible 'was not written to teach us astronomy.'

Pope Paul V, who was disturbed by the discussion of biblical interpretation, at the time a serious issue with the Protestants, appointed a commission to study the question of the earth's motion. The commission decided that the teachings of Copernicus were probably contrary to the Bible, and on 28 February 1616 Galileo was told by Cardinal Bellarmine that he could no longer hold or defend the heliocentric theory. On March 3 Bellarmine reported that Galileo had acquiesced to the Pope's warning, and that ended the matter for the time being.

On 5 March 1616 the Holy Office of the Inquisition in Rome placed the works of Copernicus and all other writings that supported it on the Index, the list of books that Catholics were forbidden to read, including those of Kepler. The decree held that believing the sun to be the immovable centre of the world is foolish and absurd, philosophically false and formally heretical. Pope Paul V instructed Cardinal Bellarmine to censure Galileo, admonishing him not to hold or defend Copernican doctrines any longer.

After his censure Galileo returned to his villa at Arcetri outside Florence, where for the next seven years he remained silent. But then in 1623, after the death of Paul V, Galileo took hope when he learned that his friend Maffeo Cardinal Barbarini had succeeded as Pope Urban VIII. Heartened by his friend's election, Galileo proceeded to publish a treatise entitled *Il Saggiatore* (The Assayer), which appeared later that year, dedicated to Urban VIII.

Il Saggiatore grew out of a dispute over the nature of comets between Galileo and Father Horatio Grassi, a Jesuit astronomer. This had been stimulated by the appearance in 1618 of a succession of three comets, the third and brightest of which remained visible until January 1619. Grassi, writing anonymously, took the view of Tycho Brahe that the comets were phenomena occurring in the celestial regions, using this as an argument against the Copernican theory in favour of the Tychonic model. Galileo, who was bedridden at the time, discussed the comets with his disciple Mario Giudecci, who gave two lectures on the subject at the Florentine Academy. The lectures were published in Giudecci's name, though in his introductory remarks he acknowledged that the ideas he presented were largely those of Galileo. In these lectures the ideas of the anonymous Jesuit were criticized, particularly concerning his support of the Tychonic model rather than the Copernican theory. Grassi, writing under the pseudonym of Lotario Sarsi, replied with a direct attack on Galileo's views in a book, published later in 1619. After a long delay, Galileo responded with *Il Saggiatore*, in the form of a letter to his disciple Virginio Cesarini, published by the Accademia dei Lincei in 1623.

Since Galileo could no longer defend the Copernican theory, in *Il Saggiatore* he avoided the contentious question of the earth's motion, but rather presented a general scientific approach to the study of celestial phenomena. Responding to the charge that he had criticized the ideas of an established authority like Tycho Brahe, he replies, 'In Sarsi I seem to discern the firm belief that in philosophizing one must support oneself upon the opinion of some celebrated author, as if our minds ought to remain completely sterile and barren unless wedded to the reasoning of some other person.' He goes on to say that philosophy is not a book of fiction such as the *Iliad* or *Orlando Furioso*, where 'the least important thing is whether what is written is true,' after which he makes his famous statement that the book of nature is written in mathematical characters.

> Well, Sarsi, that is not how matters stand. Philosophy is written in this grand book, the universe, which stands continually open to our gaze. But the book cannot be understood unless one learns to comprehend the language and read the letters in which it is composed. It is written in the language of mathematics, and its characters are triangles, circles and other geometrical figures, without which it is humanly impossible to understand a single word of it, without these, one wanders about it in a dark labyrinth.

Il Saggiatore was favourably received in the Vatican, and Galileo went to Rome in the spring of 1623 and had six audiences with the Pope. Urban praised the book, but he refused to rescind the 1616 edict against the Copernican theory, though he said that if it had been up to him the ban would not have been imposed. Galileo did receive Urban's permission to discuss Copernicanism in a book, but only if the Aristotelian-Ptolemaic model was given equal and impartial attention.

Encouraged by his conversations with Urban, Galileo spent the next six years writing a book in Italian whose title in English is *Dialogue*

Concerning the Chief World Systems Ptolemaic and Copernican, which was completed in 1630 and finally published in February 1632. The book is divided into four days of conversations between three friends: Salviati, a Copernican, who is a spokesman for Galileo himself; Simplicio, an Aristotelian, named for the famous commentator on Aristotle, Simplicius; and Sagredo, an educated layman, named for Galileo's late friend, Giovan Francesco Sagredo, whom each of the other two tries to convert to his point of view.

The first day is devoted to a critical examination of Aristotelian cosmology, particularly those that have to do with the distinction between the terrestrial and celestial regions, and the notion that the earth is the immobile centre of the universe, both of which are rejected. However, the Aristotelian theory that the natural motion of celestial bodies is circular was accepted, but Galileo argued that circular motion is also natural for bodies on the rotating earth. He also rejected the Aristotelian notion of motion as a process requiring a continuous cause, and instead he gave a definition that would enable him to measure it, saying 'Let us call velocities equal, when the spaces passed have the same proportion as the times in which they are passed.'

On the second day the objections against the earth's motions are refuted on physical grounds. Many of Galileo's arguments here, though persuasive, are based on his erroneous notion that circular motion is natural for bodies on the rotating earth. Galileo does say in his preface that any experiment performed would give the same result whether the earth is in motion or at rest. Concerning the Aristotelian theory of gravitation, Galileo has Sagredo give this response to Simplicio's statement that everyone knows what causes bodies to fall downwards is gravity:

> You are wrong, Simplicio; you should say that everyone knows that it is called gravity. But I am not asking you for the name but the essence of the thing. Of this you know not a bit more than you know the essence of the stars in gyration [...] We don't really understand what principle or what power it is that moves a stone downwards, any more than we understand what moves it upward after it has left the projector, or what moves the moon round. We have merely, as I said, assigned to the first the more specific and definite name *gravity*, whereas to the second we assign the more general term impressed power (*vita impressa*), and the last we call an *intelligence*, either *assisting* or informing, and as the cause of infinite other motions we give *nature*.

Debating the Copernican and Ptolemaic Models

The third day is concerned with the yearly motion of the earth around the sun, presenting arguments for and against Copernicanism. Galileo has Salviati praise Aristarchus and Copernicus for proposing a theory that they arrived at through reasoning though it seems to be contrary to sensory evidence.

> Nor can I sufficiently admire the eminence of those men's intelligence who have received and held it [the heliocentric theory] to be true, and with the sprightliness of their judgements have done such violence to their own senses, that they have been able to prefer that which their reason dictated to them to that which sensible experiences represented to the contrary [...] I cannot find any bounds for my admiration how reason was able, in Aristarchus and Copernicus, to commit such a rape upon their senses as in spite of them, to make herself mistress of her belief.

Galileo describes several celestial phenomena that seem to support the Copernican theory 'until it might seem that this must triumph absolutely,' but he goes on to say that these arguments will only simplify astronomy and will not show 'any necessity imposed by nature.' Here, in comparing the two world systems, Galileo is often unfair in his criticism and exaggerates his claims for the superiority of the heliocentric theory.

The fourth day is devoted to Galileo's erroneous theory of tidal action, which he attributed to the three-fold Copernican motions of the earth, and which he believed to be conclusive proof of the earth's rotation.

Despite these defects, the arguments for Copernicanism were very persuasive and poor Simplicio, the Aristotelian, is defeated at every turn. Simplicio's closing remark represents Galileo's attempt to reserve judgment in the debate, where he says that 'it would still be excessive boldness for anyone to limit and restrict the Divine power and wisdom to some particular fancy of his own.' This statement apparently was almost a direct quote of what Pope Urban had said to Galileo in 1623. When Urban read the *Dialogue* he remembered these words and was deeply offended, feeling that Galileo had made a fool of him and taken advantage of their friendship to violate the 1616 edict against teaching Copernicanism. The Florentine ambassador Francesco Niccolini reported that after discussing the *Dialogue* with Urban, the Pope broke out in great anger and fairly shouted, 'Your Galileo has ventured to meddle with things that he ought not, and with the most grave and dangerous subjects that can be stirred up these days.'

Urban directed the Holy Office to consider the affair and summoned Galileo to Rome. Galileo arrived in Rome in February 1633, but his trial before the court of the Inquisition did not begin until April. There he was accused of having ignored the 1616 edict of the Holy Office not to teach Copernicanism. The court deliberated until June before giving its verdict, and in the interim Galileo was confined in the palace of the Florentine ambassador. He was then brought once again to the Holy Office, where he was persuaded to acknowledge that he had gone too far in his support of the Copernican 'heresy,' which he now abjured. His abjuration was according to the prescribed formula: 'I, Galileo, son of the late Vicenzio Galilei of Florence, seventy years of age [...] abandon completely the false opinion that the sun is at the center of the world and does not move and that the earth is not the center of the world and moves.'

Galileo was thereupon sentenced to indefinite imprisonment and his *Dialogue* placed on the Index. The sentence of imprisonment was immediately commuted to allow him to be confined in one of the Roman residences of the Medici family, after which he was moved to Siena and then, in April 1634, allowed to return to his villa at Arcetri.

After he returned home Galileo took up again the researches he had abandoned a quarter of a century earlier, principally the study of motion. This gave rise to the last and greatest of his works, *Discourses and Mechanical Demonstrations Concerning Two New Sciences, of Mechanics and of Motions*, which he dictated to his disciple Vincenzo Viviani. The work was completed in 1636, when Galileo was 72 and suffering from failing eyesight. Since publication in Italy was out of the question because of the papal ban on Galileo's works, his manuscript was smuggled to Leyden, where the *Discourses* was published in 1638, by which time he was completely blind.

The *Discourses* is organized in the same manner as the *Dialogues*, divided into four days of discussions among three friends. The first day is devoted to subjects that Galileo had not resolved to his satisfaction, particularly his speculations on the atomic theory of matter. The second day is taken up with one of the two new sciences, now known in mechanical engineering studies as 'strength of materials.'

The third and fourth days are devoted to the second of the two new sciences, kinematics, the mathematical description of motion, including motion at constant velocity, uniformly accelerated motion as in free fall, non-uniformly accelerated motion as in the oscillation of a pendulum, and two-dimensional motion as in the path of a projectile.

The section 'On Local Motion' discussed on the third day, presents Galileo's definition of uniformly accelerated motion as a motion in which, 'when starting from rest, acquires during equal time intervals equal increments of velocity.' He says he adopted this because Nature employs 'only those means which are most common, simple and easy.'

This definition is equivalent to the modern equation $v = at$, where v is the velocity, a is the constant acceleration and t is the time. If a is, for example, 2 foot per second, then after 1 second $v = 2$, after 2 seconds $v = 4$, after 3 seconds $v = 6$, etc., all in units of feet per second. Using graphical analysis, derived from Oresme, Galileo showed that s, the distance traversed, increases each second as 1, 3, 5 and so on, which is equivalent to the modern equation $s = \frac{1}{2} at^2$. These two equations, $v =$ and $s = \frac{1}{2} at^2$, are the basic laws of modern kinematics for the case of uniformly accelerated motion starting from rest. This is the first statement of the laws of kinematics, the purely mathematical description of motion, independent of any dynamical theory. As Salviati says regarding Galileo's purpose in formulating these laws:

> At present it is the purpose of our Author merely to investigate and to demonstrate some of the properties of accelerated motion (whatever the cause of the acceleration may be – meaning that the momentum of its velocity goes on increasing after proportion from rest in simple proportionality to time, which is the same as saying that in equal time intervals the body receives equal increments of velocity; and if we find that the properties of [accelerated motion] which will be demonstrated are realized are realized in freely falling and accelerated bodies, we may conclude that the assumed definition includes such a motion of heavy bodies and that their speed goes on increasing as the time and the duration of the motion.

On the third day Galileo showed that the path of a projectile was the combination of two motions, one of them at constant velocity in the horizontal direction and the other with constant acceleration vertically downward, the resultant of which was a parabola. He also showed that the range of a projectile on a horizontal plane was greatest when the angle of elevation was 45°, and that for angles smaller and larger than this by a given amount the ranges will be equal to one another. Salviati remarks that this showed the superiority of a theoretician able to predict previously unnoticed results, as compared to a purely empirical approach:

> The knowledge of a single fact acquired through the discovery of its causes prepares the mind to ascertain and understand other facts without need of recourse to experiment, precisely as in the present case, where by argument alone the Author proves with certainty that the maximum range occurs when the elevation is 45°. He thus demonstrates what perhaps has never been observed in experienced, namely, that of other shots those which exceed or fall short of 45° by equal amounts have equal ranges.

Galileo died at Arcetri on 8 January 1642, 38 days before what would have been his 78th birthday. The Grand Duke of Tuscany sought to erect a monument in his memory, but he was advised not to do so for fear of giving offence to the Holy Office, since the Pope had said that Galileo 'had altogether given rise to the greatest scandal throughout Christendom.'

After Galileo's death a note in his hand was found on the preliminary leaves of his own copy of the *Dialogues*, which he probably wrote after the Holy Office imprisoned him for supporting the Copernican 'heresy.'

> Take note theologians, that in your desire to make matters of faith and of proposition relating to the fixity of sun and earth you may run the risk of eventually having to condemn as heretics those who would decide the earth to stand still and the sun to change position – eventually, I say – at such a time as it might be physically or logically proved that the earth moves and the sun stands still.

CHAPTER 15
THE NEWTONIAN SYNTHESIS

The pioneering observations and new theories of Copernicus, Tycho Brahe, Kepler and Galileo, together with those of some of their contemporaries, represent part of an intellectual upheaval that came to be called the Scientific Revolution, which continued through the seventeenth century and on into the early years of the eighteenth, a period during which the world view of Western Europe changed profoundly and modern scientific culture emerged.

One particularly influential system of natural philosophy that emerged in the seventeenth century was mechanism, which held that all natural phenomena was due to one single kind of change, the motion of matter. The approach of the mechanistic philosophy of nature is broadly summarized by Pierre Gassendi (1592–65), a Catholic priest who in 1647 published a work in which he attempted to reconcile the atomic theory with Christian doctrine. According to Gassendi: 'There is no effect without a cause; no cause acts without motion, nothing acts on distant things except through itself or an organ or transmission; nothing moves unless it is touched, whether directly or through an organ or through another body.'

Gassendi's mechanism was based on the atomic theory as interpreted by Lucretius, in which physical properties are traced to the imagined size and shape of the component particles. Gassendi was deeply influenced by Isaac Beeckman (1566–1637), who expressed his corpuscular form of mechanism in the statement that 'all properties arise from

[the] motion, shape and size [of the fundamental particles]. So that each of these three things must be considered.'

A number of different approaches to scientific investigation were formulated in the seventeenth century. One was the empirical, inductive method proposed by Francis Bacon (1561–1626); another was the theoretical, deductive approach of René Descartes (1596–1650).

According to Bacon, the new science should be based primarily on observation and experiment, and it should arrive at general laws only after a careful and thorough study of nature. In his *Novum Organum*, published in 1620, Bacon criticized the existing state of scientific knowledge. 'The subtlety of nature greatly exceeds that of sense and understanding, so that those fine meditations, speculations and fabrications of mankind are unsound, but there is no one to stand by and point it out. And just as the sciences we now have are useless for making discoveries of practical use, so the present logic is useless for the discovery of the sciences.'

Bacon never accepted the Copernican theory, which he called a 'hypothesis,' and he criticized both Ptolemy and Copernicus for presenting nothing more than 'calculations and predictions' rather than 'philosophy [...] what is found in nature herself, and is actually and really true.'

Descartes sought to give physical laws the same certitude as those of mathematics. As he wrote in a letter to Marin Mersenne: 'In physics I should consider that I knew nothing if I were able to explain only how things might be, without demonstrating that they could not be otherwise. For having reduced physics to mathematics, this is something possible, and I think that I can do it within the small compass of my knowledge, though I have not done it in my essays.'

Whereas in philosophy Descartes began with the existence of the self (*Cogito ergo sum*, I am thinking, therefore I exist), in physics he started with the existence of matter, its extension in space, and its motion through space. That is, everything in nature can be reduced to matter in motion. Matter exists in discrete particles which collide with one another in their ceaseless motions, changing their individual velocities in the process, but with the total 'quantity of motion' in the universe remaining constant. Descartes writes of the divine origin of this law in his *Principles of Philosophy* (1644), an extraordinarily detailed and elaborate model of the physical universe based on his corpuscular mechanistic theory. Speaking of God, he says that 'In the beginning, in his omnipotence, he created matter, along with its motion and rest, and now, merely by his regular concurrence, he preserves the same amount of motion and rest in the material universe as he put there in the beginning.'

Descartes presented his method in *Rules for the Direction of the Mind*, completed in 1628 but not published until after his death, and in the *Discourse on Method*, published in 1637 along with appendices entitled *Optics*, *Geometry* and *Meteorology*. He gave the final form of his three laws of nature in *The Principles of Philosophy* (1644). The first law, the principle of inertia, states that 'Each and every thing, insofar as it can, always continues in the same state, and thus what is once in motion always continues to move.' The second law states that 'all motion is in itself rectilinear [...] every piece of matter, considered in itself, always tends to continue moving, not in any oblique path but only in a straight line.' The third law is concerned with collisions: 'if a body collides with another body that is stronger than itself, it loses none of its motion; but if it collides with a weaker body, it loses a quantity of motion equal to that which it imparts to the other body.'

The *Optics* presents Descartes's mechanistic theory of light, which he conceived of as a series of impulses propagated through the finely dispersed micro-particles that fill the spaces between macroscopic bodies, leaving no intervening vacuum. This model gave him the right form for the law of refraction, but in his derivation he took the velocity of light to be greater in water than in air, which is not true.

The *Geometry* was inspired by what Descartes called the 'true mathematics' of the ancient Greeks, particularly Pappus and Diophantus. Here he provided a geometric basis for algebraic operations, which to some extent had already been done by his predecessors as far back as al-Khwarizmi. The symbolic notation used by Descartes quickly produced great progress in algebra and other branches of mathematics. His work gave rise to the branch of mathematics now known as analytic geometry, which had been anticipated by Pierre Fermat (1539–65). Fermat, inspired by Diophantus and Apollonius, was also one of the founders of modern number theory and probability theory.

Descartes's *Meteorology* includes his model of the rainbow, in which he used the laws of reflection and refraction to obtain the correct values of the angles at which the primary and secondary bows appear. He begins his explanation by pointing out that rainbows occur not only in the sky but also in illuminated fountains and sprays, so that it is not solely a celestial phenomenon but rather one involving light and individual drops of water. He tested this hypothesis by taking a spherical glass flask full of water, holding it up at arm's length in the sunlight and moving it up and down so that colours are produced.

He concluded, as had Dietrich of Freiburg, that the primary rainbow was produced by two refractions and one internal refection within each of the raindrops, while the secondary bow was generated by two refractions and two internal reflections, the second of which had the effect of inverting the spectrum.

He then did several experiments to show that the actual dispersion of light into colours was due only to refraction and not reflection. The explanation that he proposed was an extremely detailed mechanistic model based on his micro-corpuscular theory of matter. One of the assumptions that Descartes had made in this theory is that light travels more rapidly in dense media such as water and glass than in air, which is incorrect.

Chapters 8 through 12 of Descartes's *Le Monde* present his mechanistic cosmology, based on his theory of matter and laws of motion. This hypothetical 'new world' that he describes consisted of an indefinite number of contiguous vortices, each with a star at its centre. He argued that the stars were the sources of light just like our sun, for 'if we consider how bright and glittering the rays of the fixed stars are, despite the fact that they are an immense distance from the sun, we will not find it hard to accept that they are not like the sun. Thus if we are as close to one of them as we are to the sun, that star would in all probability appear as large and luminous as the sun.' He held that each of these stars is the centre of a planetary system, all carried around by the motion of the particles of the three types of matter that he believed filled all of space.

Descartes's vortex theory was generally accepted at first, but the researches of Christiaan Huygens (1629–95) showed conclusively that it was completely incorrect. Huygens was led to his rejection of the vortex theory by his studies of dynamics. In one of his studies he considered a situation in which a lead ball is attached to a string held by a man standing at the centre of a rotating platform. When the platform rotates the man feels an outward or centrifugal force in the string attached to the ball, which in turn experiences an inward or centripetal force due to the string. Huygens found that the centripetal force on the ball was directly proportional to the mass of the ball and the square of its velocity, and inversely proportional to the radius of its circular path, thus establishing the basis of dynamics for circular motion. This and his researches on the laws of collisions were what led Huygens to conclude that the Cartesian cosmology was in error. As he said in 1693, he could find 'almost nothing I can approve as true in all the physics and metaphysics' of Descartes.

Huygens found that Descartes's rules concerning collisions were not mutually consistent. Descartes believed in both relativity of motion and conservation of motion, while Huygens realized that these are incompatible with one another. He saw that his task was to clarify just what relativity of motion implied for collisions.

In his *De motu corporum ex percussione*, the first version of which was completed in the mid-1550s, Huygens, by looking at the same phenomenon in two different frames of reference, showed that the centre of gravity of a system is unchanged in an elastic collision.

Huygens also did pioneering work on motion in his treatise on the pendulum clock, the *Horologium oscillilatorium*, published in 1673. The thesis describes an isochronous pendulum clock invented by Huygens in 1656, in which the pendulum bob swings against a cycloidal surface, which makes its period independent of amplitude for all angles. Huygens describes the significance of his research, noting that 'the simple pendulum does not naturally provide an accurate and equal measure of time since the wider motions are observed to be slower than the narrower ones. But by a geometrical method we have found a different and previously unknown way to suspend the pendulum; and we have discovered a line whose curvature is marvelously and quite rationally suited to give the required equality to the pendulum.'

Part 2 of the *Horologium* begins with three hypotheses on dynamics. The first, which is a clear statement of the principle of inertia, states that in the absence of gravity a body will continue in any motion it already has in a straight line at constant velocity. The second hypothesis is that gravity always acts so as to impose a downward component on any uniform motion the body has, and the third says that these motions are independent of one another.

The Aristotelian notion that a vacuum was impossible was shown to be incorrect by several seventeenth-century scientists, beginning with

Evangelista Torricelli (1608–47) and Blaise Pascal (1623–62). Torricelli's invention of the barometer in 1643 led him to conclude that the closed space above the mercury column represented at least a partial vacuum, and that the difference in the height of the two columns in the U-tube was a measure of the weight of a column of air extending to the top of the atmosphere. Pascal had a barometer taken to the top of the Puy de Dôme, a peak in central France, and it was observed that the difference in height of the two columns was less than at sea-level, verifying Torricelli's conclusions. The results of this experiment led Pascal to urge all disciples of Aristotle to see if the writings of their master could explain the results. 'Otherwise,' he wrote, 'let them recognize that experiments are the real masters that we should follow in physics; that the experiment done in the mountains overturns the universal belief that nature abhors a vacuum.'

The German engineer Otto von Guericke (1602–80) discovered that it was possible to pump air as if it were water, allowing him to produce a vacuum mechanically. In a famous experiment at Magdeburg in 1657, he pumped the air out of a spherical cavity made by fitting together two copper hemispheres, and showed that the resulting differential pressure was so great that not even two teams of horses pulling in opposite directions could force the two halves of the sphere apart.

Guericke's demonstration led the Irish chemist Robert Boyle (1627–91) to have a vacuum pump constructed by the instrument maker Ralph Greatorex. The design of the pump was subsequently improved by the English physicist Robert Hooke (1635–1703). Boyle fitted the pump with a Torricellian barometer and noted the change in the level of mercury as the tube was evacuated. He then used the pump to do research on pneumatics, which he published in 1660 under the title *New Experiments Physico Mechanical, Touching the Spring of Air and its Effects*. His conclusions were that a vacuum can be produced, or at least a partial one; that sound does not propagate in a vacuum; and that air is necessary for life or a flame. He also concluded that air is an elastic fluid that exerts a pressure against whatever restricts it and expands when it is relieved of external constraints: 'Air ether consists of, or at least abounds with, parts of such a nature, that in case they be bent or compress'd by the incumbent part of thermosphere, they do endeavour, as much as in them lies, to free themselves from that pressure, by bearing upon the contiguous bodies that keep them bent.' In an appendix to the second edition of this work, published in 1662, he established the relationship now known as Boyle's Law, that the pressure exerted by a gas is inversely proportional to its volume.

Boyle was influenced by both Francis Bacon's empiricism and Descartes's mechanistic view of nature. He was also influenced by the natural philosophy of Epicurus, revived by Pierre Gassendi. This led Boyle to adopt a divinely ordained corpuscular version of mechanism, which he described in his treatise on *Some Thoughts about The Excellence and Grounds of the Mechanical Philosophy*, published in 1674. As he concluded concerning the universality of mechanism: 'By this very thing that the mechanical principles are so universal, and therefore applicable to so many other things, they are rather fitted to include, than necessitated to exclude, any other hypothesis, that is founded in nature, as far as it is so.'

The culmination of the scientific developments that had had taken place from the time of Copernicus through that of Galileo came with the career of Isaac Newton (1642–1727), whose supreme genius made him the central figure in the emergence of modern science.

Newton was born on 25 December 1642, the same year that Galileo had died. His birthplace was the manor house of Woolsthorpe in Lincolnshire, England. His father, an illiterate farmer, had died three months before Isaac was born, and his mother remarried three years later, though she was widowed again after eight years. When Newton was 12 he was enrolled in the grammar school at the nearby village of Grantham, and he studied there until he was 18. His maternal uncle, a Cambridge graduate, sensed that his nephew was gifted and persuaded Isaac's mother to send the boy to Cambridge, where he was enrolled at Trinity College in June 1661.

At Cambridge Newton was introduced to both Aristotelian science and cosmology as well as the new physics, astronomy and mathematics of Copernicus, Kepler, Galileo, Fermat, Descartes, Huygens and Boyle. In 1663 he began studying under Isaac Barrow (1630–77), the newly appointed Lucasian professor of mathematics and natural philosophy. Barrow edited the works of Euclid, Archimedes and Apollonius, and published his own works on geometry and optics, with the assistance of Newton.

By Newton's own testimony he began his researches in mathematics and physics late in 1664, shortly before an outbreak of plague closed the university at Cambridge and forced him to return home. During the next two years, his *anni mirabiles*, he says that he began his researches in the calculus and the dispersion of light, and discovered his law of universal gravitation and motion as well as the concepts of centripetal force and acceleration.

> In the beginning of the year 1665 I found the Method of approximating series & the Rule for reducing any dignity of any Binomial into such a series. The same year in May I found the method of Tangents of Gregory & Slusius, and in November had the direct method of fluxions & the next year in January had the Theory of Colours & in May following I had entrance into ye inverse method of fluxions. And the same year I began to think of gravity extending to ye orb of the Moon & (having found out how to estimate the force with which [a] globe revolving within a sphere presses the surface of the sphere) from Kepler's rule of the periodic times of the Planets being sesquialternate proportion of their distances from the center of their Orbs. I deduced that the forces which keep the Planets in their Orbs must [be] reciprocally as the squares of their distances from the centers about which they revolve: and thereby compared the force required to keep the Moon in her Orb with the force of gravity at the surface of the earth & found them answer pretty nearly. All this was in the two plague years 1665 and 1666 for in those years I was in the prime of my age for invention & minded Mathematicks & Philosophy more than at any time since.

This indicates that Newton had derived the law for centripetal force and acceleration by 1666, some seven years before Huygens, though he did not publish it at the time. He applied the law to compute the centripetal acceleration at the earth's surface caused by its diurnal rotation, finding that it was less than the acceleration due to gravity by a factor of 250, thus settling the old question of why objects are not flung off the planet by its rotation. He computed the centripetal force necessary to keep the moon in orbit, comparing it to the acceleration due to gravity at the earth's surface, and found that they were inversely proportional to the squares of their distances from the centre of the earth. Then, using Kepler's third law of planetary motion together with the law of centripetal acceleration, he verified the inverse square law of gravitation for the solar system. At the same time he laid the foundations for the calculus and formulated his theory for the dispersion of white light into its component colours.

The Newtonian Synthesis

When the plague subsided Newton returned to Cambridge in the spring of 1667. Two years later he succeeded Barrow as Lucasian Professor of Mathematics and Natural Philosophy, a position he was to hold for nearly 30 years.

During the first few years after he took up his professorship Newton devoted much of his time to research in optics and mathematics. He continued his experiments on light, examining its refraction in prisms and thin glass plates as well as working out the details of his theory of colours. He built a reflecting telescope with a magnifying power of nearly 40, and then made a refractor that he claimed magnified 150 times, using it to observe planets and comets. The latter telescope came to the attention of the Royal Society, which elected him a Fellow on 11 January 1672.

As part of his obligations as a Fellow, Newton wrote a paper on his optical experiments, which he submitted on 28 February 1672, to be read at a meeting of the Society. The paper, subsequently published in the *Philosophical Transactions of the Royal Society*, described his discovery that sunlight is composed of a continuous spectrum of colours, which can be dispersed by passing light through a refracting medium such as a glass prism. He found that the 'rays which make blue are refracted more than the red,' and he concluded that sunlight is a mixture of light rays, some of which are refracted more than others. Furthermore, once sunlight is dispersed into its spectrum of colours it cannot be further decomposed. In the experiment that demonstrates this, sunlight passes through a prism and is dispersed into its component colours, with the blue rays diffracted more than the red. When the various rays passed though a second prism no further dispersion occurred, blue remained blue and red remained red. This meant that the colours seen on refraction are inherent in the light itself and are not imparted to it by the refracting medium. Newton referred to this demonstration as his *experimentum crucis*, or 'crucial experiment,' because it showed that his theory concerning the nature of light could be put on a mathematical basis, as he noted in a letter to Henry Oldenburg, Secretary of the Royal Society:

> A naturalist would scarce expect to see ye science of those [colours] become mathematicall, & yet I dare affirm that there is as much certainty in it as in any other part of Opticks. For what I shall tell concerning

them is not an Hypothesis but most rigid consequence, not conjectured by barely inferring' tis because not otherwise or because it satisfies all phenomena (the Philosophers universall Topick,) but invinced by ye mediation of experiments concluding directly & without any suspicion of doubt.

The paper was characteristic of Newton's attitude towards the approach to be followed in any scientific investigation. Later, in a controversy arising out of his first paper, Newton described his scientific method.

> For the best and safest method of philosophizing seems to be, first to enquire diligently into the properties of things, and to establish these properties by experiment, and then to proceed more slowly to hypotheses for the explanation of them. For hypotheses should be employed only in explaining the properties of things, but not assumed in determining them, unless so far as they may furnish experiments.

Ironically, the paper was widely criticized by Newton's contemporaries for just the contrary reason: that it did not confirm or deny any general philosophy of nature, and the mechanists objected that it was impossible to explain his findings on the basis of any mechanical principles. As Huygens wrote in a letter to Oldenburg:

> if it were true that from their origin some rays of light were red, others blue etc., there would remain the great difficulty of explaining by the mechanical philosophy in what this diversity of colours consist [...] for until this hypothesis has been found, [Newton] has not apprised us of what the nature of and difference between colours is, only the accident (which is certainly very considerable), of their refrangibility.

Then there were others who insisted that Newton's experimental findings were false, since they themselves could not find the phenomena that he had reported. Newton replied patiently to each of these criticisms in turn, but after a time he began to regret ever having presented his work in public.

One of those who criticized his paper was Robert Hooke, who in November 1662 was appointed as the first Curator of Experiments at the newly founded Royal Society, a position he held until his death in 1703, making many important discoveries in mechanics, optics, astronomy, technology, chemistry and geology. His lengthy critique of the paper seemed to imply that Hooke had performed all of Newton's experiments himself, while rejecting the conclusions that Newton had drawn.

The criticisms of Newton's paper led him to resign from the Royal Society early in 1673, but Oldenburg refused to accept his resignation and persuaded him to remain. Then in 1676, after a public attack by Hooke, Newton broke off almost all association with Oldenburg and the Royal Society. The following year Oldenburg died and Hooke replaced him as Secretary of the Society, whereupon he wrote a conciliatory letter in which he expressed his admiration for Newton. Referring to Newton's theory of colours, Hooke said that he was 'extremely well pleased to see those notions promoted and improved which I long since began, but had not time to compleat.'

Newton replied in an equally conciliatory tone, referring to Descartes's work on optics. 'What Descartes did was a good step. You have added much several ways, and especially in taking the colors of thin plates into philosophical consideration.'

But despite these friendly sentiments, the two were never completely reconciled, and Newton maintained his silence. Nevertheless they continued to communicate with one another, a correspondence that was to lead again and again to controversy, the bitterest dispute arising from Hooke's claim that he had discovered the law of gravitation before Newton.

Hooke wrote to Newton in November 1679, asking him 'particularly if you will let me know your thoughts of that of compounding the celestial motions of the planets of a direct motion by the tangent & an attractive motion toward the central body.' Hooke is saying that a body would normally follow a rectilinear path, and consequently if, like the planets, it is deflected from its linear motion there must be some cause, which he suggests is the attraction of the sun.

Newton replied that he had not thought about such questions for years, and had not been aware of 'your Hypotheses of compounding ye celestial motions of ye planets, of a direct motion by the tangt to ye curve.'

Hooke had proposed this notion in a paper read to the Royal Society in 1666, on 'the inflection of a direct motion into a curve by a supervening attractive principle.' He had expanded on the idea in a lecture published in 1674 entitled *Attempt to Prove the Motion of the Earth*, which

Newton appears to have in fact read. The lecture was republished in 1679 in Hooke's *Lectiones Cutlerianae*, where he describes the dynamical elements of orbital motion in his system:

> This depends upon three Suppositions. First, that all Coelestial Bodies whatsoever, have an attraction or gravitational power towards their own Centers, whereby they attract not only their own parts, and keep them from flying from them, as we may observe the earth to do, but that they do also attract all other Coelestial Bodies that are within the sphere of their activity [...] The second supposition is this, That all bodies whatsoever that are put into a direct and simple motion, will so continue to move forward in a straight line, till they are by some other effectual powers deflected and bent into a Motion, describing a Circle, Ellipse, or some other compounded Curve Line. The third supposition is, That these attractive powers are so much the more powerful in operating, by how much the nearer the body wrought upon is to their own Centers. Now what these several degrees are I have not yet experimentally verified.

By 1684 others beside Hooke and Newton were convinced that the gravitational force was responsible for holding the planets in their orbits, and that this force varied with the inverse square of their distance from the sun. Among them were the astronomer Edmund Halley (1656–1742), a good friend of Newton and a fellow member of the Royal Society. Halley made a special trip to Cambridge in August 1684 to ask Newton if he thought the Curve would be described by the planets supposing the force of attraction towards the sun to be reciprocal to the square of their distance from it. Newton replied immediately that it would be an ellipse, but he could not find the calculation, which he had done seven or eight years before. And so he was forced to rework the problem, after which he sent the solution to Halley that November.

By then Newton's interest in the problem had revived, and he developed enough material to give a course of nine lectures in the autumn term at Cambridge, under the title of *De motu corporum in gyrum* (On the Motion of Bodies in an Orbit), a treatise of nine pages that he duly gave to Halley that November. The treatise demonstrated that, for velocities below a certain limit, an inverse-square force involves an elliptic orbit with the centre of force at one focal point. Assuming certain dynamical definitions and hypotheses, Newton was also able to demonstrate Kepler's second and third laws of planetary motion.

When Halley read the manuscript of *De Motu* he realized its immense importance, and he obtained Newton's promise to send it to the Royal Society for publication. On 22 May 1686 Halley wrote to Newton saying that the Society had entrusted him with the responsibility for having the manuscript printed. But he added that Hooke, having read the manuscript, claimed that it was he who had discovered the inverse square nature of the gravitational force and thought that Newton should acknowledge this in the preface. Newton was very much disturbed by this, and in his reply to Halley he went to great lengths to show that he had discovered the inverse square law of gravitation and that Hooke had not contributed anything of consequence.

The first edition of Newton's work was published in midsummer 1687 at the expense of Halley, since the Royal Society had found itself financially unable to fund it. Newton entitled his work *Philosophicae Naturalis Principia Mathematica* (The Mathematical Principles of Natural Philosophy), referred to more simply as the *Principia*. The *Principia* begins with an ode dedicated to Newton by Halley. This is followed by a preface in which Newton outlines the scope and philosophy of his work.

> [O]ur present work sets forth mathematical principles of natural philosophy. For the basic problem of philosophy seems to be to discover the forces of nature from the phenomena of motions, and then to demonstrate the other phenomena from these forces [...] Then the motions of the planets, the comets, the moon, and the sea are deduced from these forces by propositions that are also mathematical. If only we could derive the other phenomena of nature from mechanical principles by the same kind of reasoning!

Book I begins with a series of eight definitions, of which the first five are fundamental to Newtonian dynamics. The first effectively defines 'quantity of matter,' or mass, as being proportional to the weight density times volume. The second defines 'quantity of motion,' subsequently to be called 'momentum,' as mass times velocity. In the third definition Newton says that the 'inherent force of matter,' or inertia, 'is the power of resisting by which every body, so far as it is able, perseveres in its state either of rest or of moving uniformly straight forward.' The fourth states that 'Impressed force is the action exerted upon a body to change its state either of resting or of uniformly moving straight forward.' The fifth through eighth define centripetal force as that by which bodies 'are impelled, or in any way tend, toward some point as to a center.' As an example Newton gives the gravitational force of the sun, which keeps the planets in orbit.

As regards the gravity of the earth, he refers to the example of a lead ball, projected from the top of a mountain with a given velocity, and in a direction parallel to the horizon. If the initial velocity is made larger and larger, he says, the ball will go farther and farther before it hits the ground, and may go into orbit around the earth or even escape into outer space.

The definitions are followed by a *Scholium*, a lengthy comment in which Newton gives his notions of absolute and relative time, space, place and motion. These essentially define the classical laws of relativity, which in the early twentieth century would be superceded by Einstein's theories of special and general relativity.

Next come the axioms, now known as Newton's laws of motion, three in number, each accompanied by an explanation and followed by corollaries.

> Law 1: Every body perseveres in its state of being at rest, or of moving uniformly forward, except insofar as it is compelled to change its state of motion by forces impressed [...] Law 2: A change of motion is proportional to the motive force impressed and takes place along the straight line in which that force is impressed [...] Law 3: To every action there is always an opposite and equal reaction; in other words, the action of two bodies upon each other are always equal, and always opposite in direction.

The first law is the principal of inertia, which is actually a special case of the second law when the net force is zero. The form used today for the second law is that the force F acting on a body is equal to the time rate of change of the momentum p, where p equals the mass m times the velocity v; if the mass is constant then $F = ma$, a being the acceleration, the time rate of change of the velocity. The third law says that when two bodies interact the forces they exert on one another are equal in magnitude and opposite in direction.

The introductory section of the *Principia* is followed by Book I, entitled 'The Motion of Bodies.' This begins with an analysis of motion in general essentially using the calculus. First Newton analysed the relations between orbits and central forces of various kinds. From this he was able to show that if and only if the force of attraction varies as the inverse square of the distance from the centre of force then the orbit is an ellipse, with the centre of attraction at one focal point, thus proving Kepler's second law of motion. Elsewhere in Book I he proved Kepler's first and third laws. He also used his third law of motion to deal with problems involving two bodies mutually attracting one another, where he notes that neither of the two bodies can be considered to be at rest: 'For attractions are always directed toward bodies, and – by the third law – the actions of attracting and attracted bodies are always mutual and equal; so that if there are two bodies, neither the attracting and attracted body can be at rest, but both [...] revolve around a common center of gravity as if by mutual attraction.'

Book II is also entitled 'The Motion of Bodies,' for the most part dealing with forces of resistance to motion in various type of fluids. One of Newton's purposes in this analysis was to see what effect the hypothetical aether in Descartes's cosmology would have on the motion of the planets. His studies showed that the Cartesian vortex theory was completely erroneous, for it ran counter to the laws of motion in resisting media that he established in Book II of the *Principia*.

The third and final book of the *Principia* is entitled 'The System of the World,' beginning with three 'Rules for the Study of Natural Philosophy.' After this comes a section on 'Phenomena,' 6 in number, followed by 42 propositions, each accompanied by a theorem and sometimes followed by a *Scholium*. This is in turn followed by a general *Scholium* and a concluding section entitled 'The System of the World.'

The six phenomena concern the motion of the planets and the earth's moon, along with observations concerning Kepler's second and third laws of planetary motion. He concludes that the planets, 'by radii drawn to the centre [...], describe areas proportional to the times, and their periodic times – the fixed stars being at rest – are as the 3/2 powers of their distances from that center.'

The first six propositions are arguments to show that the inverse-square gravitational force explains the motion of the planets orbiting the sun, the satellites of Jupiter, and the earth's moon, as well as the local gravity on the earth itself. The seventh proposition states Newton's law of universal gravitation: 'Gravity exists in all bodies universally and is proportional to the quantity of matter in each.'

Proposition 13 states Kepler's first and second laws of planetary motion: 'The planets move in ellipses that have a focus in the centre of the sun, and by radii drawn to that centre they describe areas proportional to the times.'

Proposition 18 claims 'The axes of the planets are smaller than the diameters that are drawn perpendicular to the axes,' that is, the planets are oblate spheres. Newton correctly attributed this effect to the centrifugal forces arising from the axial rotation of the planets, so that the earth, for example, is flattened at the poles and bulges around the equator.

Proposition 24 presents Newton's theory of tidal action, that 'the ebb and flow of the sea arises from the actions of the Sun and Moon,' finally solving a problem that dated back to the time of Aristotle.

Proposition 39 is 'To find the precession of the equinoxes,' including the gravitational forces of both the sun and the moon on the earth. Newton correctly computed that 'the precession of the equinoxes is more or less 50 seconds [of arc] annually,' thus solving another problem that had preoccupied astronomers for some 2,000 years.

Lemma 4 states that 'The comets are higher than the Moon, and move in the planetary regions.' In the lemmas and propositions that follow, Newton discusses the motion of comets, showing that they move in elliptical orbits around the sun, thus reappearing periodically, as did the one known as Halley's comet, which had been observed in 1682 after disappearing 75 years before. He also speculated on the nature of comets, saying, as had Kepler, that the tail of a comet represents vaporization from the comet's head as it approaches the sun.

This is followed by a general *Scholium*, in which Newton explains that mechanism alone cannot explain the universe, whose harmonious order indicated to him the design of a Supreme Being. 'This most elegant system of the sun, planets, and comets could not have arisen without the design and dominion of an intelligent and powerful being.'

A second edition of the *Principia* was published in 1713 and a third in 1726, in both cases with a preface written by Newton. Meanwhile Newton had in 1704 published his researches on light, much of which had been done early in his career. Unlike the *Principia*, which was in

Latin, the first edition of his new work was in English, entitled *Opticks, or a Treatise of the Reflexions, Refractions, Inflexions and Colours of Light*. The first Latin edition appeared in 1706, and subsequent English editions appeared in 1717/1718, 1721 and 1730; the last, which came out three years after Newton's death, bore a note stating that it was 'corrected by the author's own hand, and left before his death, with his bookseller.'

Like the *Principia*, the *Opticks* is divided into three Books. At the very beginning of Book I Newton reveals the purpose he had in mind when composing his work. 'My design in this Book,' he writes, 'is not to explain the Properties of Light by Hypotheses, but to propose and prove them by Reason and Experiment.'

The topics dealt with in Book I include the laws of reflection and refraction, the formation of images, and the dispersion of light into its component colours by a glass prism. Other topics include the properties of lenses and Newton's reflecting telescope; the optics of human vision; the theory of the rainbow; and an exhaustive study of colour. Newton's proof of the law of refraction is based on the erroneous notion that light travels more rapidly in glass than in air, the same error that Descartes had made. This error stems from the fact that both of them thought that light was corpuscular in nature.

Newton's corpuscular view of light stemmed from his acceptance of the atomic theory. He writes of his admiration for 'the oldest and most celebrated Philosophers of Greece [...] who made a Vacuum, and Atoms, and the Gravity of Atoms, the first Principles of their Philosophy [...] All these things being consider'd, it seems to me that God in the Beginning formed Matter in solid, hard, impenetrable, moveable Particles, of such Sizes and Figures, and with such other Properties and in such Proportions to Space, as much conduced to the End for which he had form'd them.'

Book II begins with a section entitled 'Observations concerning the Reflexions, Refractions, and Colours of thin transparent bodies.' The effects that he studied here are now known as interference phenomena, where Newton's observations are the first evidence for the wavelike nature of light.

In Book II Newton also comments on the work of the Danish astronomer Olaus Roemer (1644–1710), who in 1676 measured the velocity of light by observing the time delays in successive eclipses of the Jovian moon Io as Jupiter receded from the earth. Roemer's value for the velocity of light was about a fourth lower than the currently accepted one of slightly less than 300,000 kilometres per second, but it was nevertheless the first measurement to give an order of magnitude estimation of one of the fundamental constants of nature. Roemer computed that light would take 11 minutes to travel from the sun to the earth, as compared to the correct value of eight minutes and twenty seconds. Newton seems to have made a better estimate of the speed of light than Roemer, for in Book II of the *Opticks* he says that 'Light is propagated from luminous Bodies in time, and spends about seven or eight Minutes of an Hour in passing from the Sun to the Earth.'

In Book III the opening section deals with Newton's experiments on diffraction. The remainder of the book consists of a number of hypotheses, not only on light, but on a wide variety of topics in physics and philosophy. The first edition of the *Opticks* had 16 of these Queries, the second 23, the third and fourth 31. It would seem that Newton, in the twilight of his career, was bringing out into the open some of his previously undisclosed speculations, his heritage for those who would follow him in the study of nature.

Meanwhile Newton had been involved in a dispute with the great German mathematician and philosopher Gottfried Wilhelm Leibnitz (1646–1716), the point of contention being which of them had been the first to develop the calculus. According to his own account, Newton first conceived the idea of his 'method of fluxions' around 1665–66, although he did not publish it until 1687, when he used it in the *Principia*. He first published his work on the calculus independently in a treatise that came out in 1711. Leibnitz began to develop the general methods of the calculus in 1675, though he did not publish his work until 1684. The version of calculus formulated by Leibnitz, whose notation was much like that used today, caught on more rapidly than that of Newton, particularly on the continent. Newton's bitterness over the dispute was such that in the third edition of the *Principia* he deleted all reference to Leibnitz, who until the end of his days continued to accuse his adversary of plagiarism.

Aside from his work in science, Newton also devoted much of his time to studies in alchemy, prophecy, theology, mythology, chronology and history. His most important non-scientific work is *Observations upon the Prophecies of Daniel, and the Apocalypse of St John*, which is considered to be a possible key to the method of his alchemical studies, as evidenced by such notions as his analogy between the 'four metals' of alchemy and the four beasts of the apocalypse.

In 1689 Newton was elected by the constituency of Cambridge University to serve as Member of Parliament. He was made Warden of the Mint in March 1696, whereupon he appointed William Wiston as his deputy in the Lucasian professorship at Cambridge. He finally resigned his professorship on 10 March 1710, shortly after his second election as MP for the university. He was knighted by Queen Anne at Trinity College on 16 April 1705; on 17 May 1706 he was defeated in his third campaign for the university's seat in Parliament.

Newton died in London on 20 March 1727, four days after presiding over a meeting of the Royal Society, of which he had been President since 1703. His body lay in state until 4 April, when he was buried with great pomp in Westminster Abbey. The baroque monument marking his tomb shows Newton in a reclining position, along with a weeping female figure, representing Astronomy, Queen of the Sciences, sitting on a globe above. The inscription on the tomb concludes in stating that 'Let Mortals rejoice That there has existed such and so great an Ornament to the Human Race.'

His contemporaries hailed Newton's achievement as the perfection of the mechanistic philosophy of nature, and historians of the mid-twentieth century praised his work as the culmination of the Scientific Revolution. Although many historians of the new millennium now take issue with the notion of a Scientific Revolution, it is generally agreed that Newton's work culminated the long development of European science, creating a synthesis that opened the way for the scientific culture of the modern age.

Newton himself paid tribute to his predecessors when he said, in response to Hooke, 'If I have seen further than Descartes, it is by standing on the sholders of Giants.' Here he is referring to the long line of scientists, beginning with the ancient Greeks and continuing in medieval Islam and in Western Europe through the Renaissance, climaxing with the heliocentric theory of Copernicus and the new astronomy of Tycho Brahe and Kepler that led to Newton's synthesis of physics and astronomy, the climax of the Scientific Revolution that give rise to modern science.

EPILOGUE
SEARCHING FOR COPERNICUS

Much has been written about the Scientific Revolution, beginning with the publication of *De revolutionibus* in 1543, and culminating with Newton's synthesis, marked by his *Principia* in 1687 and his *Opticks* early in the following century. Many books have been written in recent years about Tycho Brahe, Kepler, Galileo, Newton and others who played prominent parts in this intellectual upheaval. But relatively little has been written about the man who began the revolution, for Copernicus spent most of his life as a secluded canon in a remote corner of the world, whereas the other notable figures of the Scientific Revolution were in public life and deeply involved in the struggle to establish the new science. They were part of a new world that was emerging in the Scientific Revolution, whereas Copernicus was still part of the older world, though he rejected the cosmology that had survived since antiquity. His only major work, *De revolutionibus*, which sparked the revolution, became obsolete after Kepler formulated his laws of planetary motion, and Copernicus himself, though venerated as the father of modern science, was reduced to the status of an icon, the man himself remaining a spectral figure.

The past three-quarters of a century have seen a revival of interest in Copernicus, climaxed by the celebration in 1973 of the 500th anniversary of his birth. The world's leading Copernican scholar in recent times was Edward Rosen, Professor of History at the City College of New York. Many of his publications on the life and work of

Copernicus and his work have been reprinted in recent years, including *Three Copernican Treatises* (1939); *Copernicus and the Scientific Revolution* (1984); *Nicholas Copernicus, Minor Works* (1985); and *Nicholas Copernicus and His Successors* (1995). The other preeminent authority is Owen Gingerich, Research Professor of astronomy and of the history of science at Harvard University and senior astronomer at the Smithsonian Astrophysical Observatory, whose many works on Copernicus include *The Eye of Heaven: Ptolemy, Copernicus, Kepler* (1993); *The Book Nobody Read, In Pursuit of the Revolutions of Nicolaus Copernicus* (2004); and, in collaboration with James MacLachlan, *Nicolaus Copernicus, Making the Earth a Planet* (2005).

I have relied heavily on the works of Rosen and Gingerich, as well as the definitive *Mathematical Astronomy in Copernicus's De revolutionibus* (1984) by N. M. Swerdlow and O. Neugebauer. Another, more recent, book that I have found to be very useful and interesting is *The First Copernican: Georg Joachim Rheticus and the Rise of the Copernican Revolution* (2006), by Dennis Danielson, Professor of English at the University of British Columbia.

The revival of interest in Copernicus has led to the creation of two museums dedicated to his life and work, one of them in Torun (Thorn), his birthplace, and the other in Frombork (Frauenburg), where he spent most of his adult life and wrote *De revolutionibus*.

The Nicolaus Copernicus Museum in Torun, one of eight divisions of the town's District Museum, is laid out in two adjoining Gothic burgher houses, one of them the birthplace of Copernicus and the other a merchant's storage space. The museum contains a collection of books and old prints from the time of Copernicus, various editions of *De revolutionibus*, models of the astronomical instruments that he used, and memorabilia associated with life in Turun in his time.

Another section of the District Museum, the Division of Polish and European Art, is housed in the Town Hall, a Gothic and Renaissance edifice, originally built in the thirteenth and fourteenth centuries and reconstructed in 1602–3. The most interesting exhibit is a portrait of Copernicus, dated *c.*1585. It shows a bust of the astronomer in the prime of his life, clad in a red sleeveless garb over a dark cassock. The portrait is by an anonymous artist, though some have attributed it to the Flemish painter and printmaker Marcus Gheeraerts (*c.*1520–*c.*1590).

There is another portrait of Copernicus in the Cathedral of St John, where the astronomer was baptized in 1473, in a side chapel dedicated to the Guardian Angels. Uninformed local belief had it that Copernicus died in Torun and was buried in the Cathedral of St John. The first monument in Torun honouring Copernicus was a symbolic tomb built in 1559 in the Chapel of the Guardian Angels. The monument has a portrait bust of Copernicus wearing a red vestment over a dark undergarment, facing slightly to the right. The background to his right shows a Crucifixion scene, while above his left shoulder there is a shelf bearing an armillary sphere and a pair of compasses.

The portrait is attributed to the Dutch painter Tobias Stimmer (1539–84), apparently based on a self-portrait by Copernicus. There is some evidence that a self-portrait of Copernicus existed during his lifetime. After his death it became the property of the Warmia chapter. When Tiedemann Giese became Bishop of Warmia in 1549 he sent a copy of this portrait to the Swiss mathematician Conrad Dasypodius (*c.*1530–1600). In 1568 Dasypodius published a work about the Copernican theory, but it is not clear whether he actually supported the heliocentric model. He also translated a work of Hero of Alexandria from Greek into Latin, probably the *Automata*. Later he designed an astronomical clock for the Strasbourg Cathedral, which was reconstructed in the years 1571–4, with a calendar dial, an astrolabe, indicators for the planets and eclipses, along with paintings, moving statues, automata and a six-tune carillon. Dasypodius hired the Swiss clockmakers Isaac and Josia Habrecht, the astronomer and musician David Wolckenstein, and the artists Tobias Stimmer and his brother Josias. The Stimmers painted large panels depicting the three Fates, Urania, Colossus, the Creation, the Resurrection, the Last Judgement and the rewards of virtue and vice, along with a figure of Copernicus based on his self-portrait. The painting of Copernicus was done by Tobias Stimmer. This was confirmed in 1873, when an inscription on its back was uncovered, reading that the Copernicus portrait was painted by Tobias Stimmer according to the original, which had been sent to Dasypodius by Doctor Tiedemann Giese from Gdansk (Danzig). This work served as the prototype for other portraits of Copernicus, including those in the Cathedral of St John and the Town Hall in Torun.

The Frombork Museum has four divisions: the Belfry, the Hospital of the Holy Ghost, the Bishop's Palace, and the Tower of Copernicus.

The Belfry commands a sweeping view of Frombork and the surrounding countryside, and has a Foucault Pendulum, with which the French physicist Jean-Bernard-Léon Foucault, in a demonstration first carried out on 3 February 1861, showed that the earth is rotating on its axis, the first direct verification of Copernicus' claim that the earth is in motion.

The Hospital of the Holy Ghost has an exhibit concerning the history of medicine, including items related to the medical work of Copernicus. It also has a collection of Warmian sculpture and painting of the seventeenth and eighteenth century.

The Bishop's Palace on Cathedral Hill features an exhibit devoted to the life and work of Copernicus, as well as another on the uses of astronomy in navigation and geodesy.

The Tower of Copernicus is in the northwestern corner of the defence walls on Cathedral Hill. Copernicus, as noted earlier, built this tower at his own expense to provide a space from which he could have a clear view of the heavens, and most of the observations that he used in *De revolutionibus* were made here. The tower has now been fitted out as if it were the astronomer's study, with examples of the instruments and other things he would have used in his researches.

Although Copernicus was buried in a crypt in Frombork cathedral, the exact location of his grave was forgotten and for over two centuries archaeologists searched for his burial place, but in vain. Then in August 2005 an archaeological team headed by Jerzy Gasowski discovered what they believed to be the remains of Copernicus. The tomb was in poor condition and part of the skeleton was missing, including the lower jaw.

A lengthy analysis was made by a team headed by Dariusz Zajdel of the Polish Police Central Forensic Laboratory, who used the skull to reconstruct a face that bore a close resemblance to the painting of Copernicus in the Torun Museum. They also found that the remains were those of a man who had died when he was about 70, the age of Copernicus at the time of his death. DNA from the skeleton matched hair samples from a book belonging to Copernicus which was preserved at the University of Uppsala in Sweden. The discovery was finally announced on 3 November 2008 by Gassowski, who said that he was virtually certain that the skeleton was that of Copernicus.

Copernicus was reburied in the same tomb in the Cathedral on 22 May 2010, which was reported in that day's edition of *The New York Times*. The requiem mass was conducted by Cardinal Jozef Kowalczyk, the newly appointed Primate of Poland. A black granite tombstone now marks the grave, with an inscription identifying Copernicus as a Warmia canon and honouring him as the founder of the heliocentric theory. Above the tombstone there is a representation of the Copernican model of the solar system, a golden sun encircled by the earth and the other five visible planets. As we have seen, this is the model described in Book I, Chapter 10 of *De revolutionibus*.

SOURCE NOTES

INTRODUCTION

p. xi 'the book that nobody read', Gingerich, *The Book Nobody Read*, p. vii

CHAPTER 1. 'THIS REMOTE CORNER OF THE EARTH'

p. 1 'this remote corner…', Copernicus, *De revolutionibus*, Preface
p. 1 'All the world…', Pastor, *History of the Popes*, vol. VI, p. 149
p. 1 'God be praised,…', ibid.
p. 7 'Thorn with its beautiful…', Repcheck, p. 35
p. 8 'pious virgin', Danielson, *The First Copernican*, p. 47

CHAPTER 2. A NEW AGE

p. 15 'open to the world…', Thucydides, ii, 39, 40
p. 15 'the true numbers' and 'true motions', Plato, *Republic*, vii, 529d
p. 15 'Let's study astronomy', ibid., vii, 530b-c
p. 16 'on what hypotheses…', Guthrie, vol. 5, p. 450
p. 19 'Heraclides supposed that…', Lloyd, *Early Greek Science*, p. 95
p. 21 'Aristarchus of Samos…', Dijksterhuis, pp. 362–3
p. 22 'on the ground…', Plutarch, xii, 923
p. 24 'in the middle of the heavens…', Ptolemy, p. 41
p. 26 'certain ancient astrologers…', Carmody, pp. 45–6
p. 27 'incorporeal motive force…', Lloyd, *Greek Science After Aristotle*, p. 159
p. 27 'was the first caliph…', Gutas, p. 30
p. 28 'It is enough…', Clagett, *Greek Science in Antiquity*, pp. 132–3
p. 28 'I shall transmit…', Minio-Paluello, 'Boethius', *DSB*, 2, 229
p. 32 'long period of study…', Crombie, *Medieval and Early Modern Science*, vol. I, p. 10
p. 33 'something new from…', ibid.
p. 33 'the opinions of the Arabs…', Haskins, *Studies in the History of Mediaeval Science*, p. 41
p. 33 'there, seeing the abundance…', Lemay, 'Gerard of Cremona', *DSB*, 15, 174

p. 34 'the fullness of Greek...', Haskins, *Studies in the History of Mediaeval Science*, p. 103
p. 34 'in order to provide Latin...', Minio-Paluello, 'William of Moerbeke', *DSB*, 9, 435

CHAPTER 3. THE JAGIELLONIAN UNIVERSITY OF KRAKOW

p. 39 'composition' and 'resolution', Crombie, *Robert Grosseteste and the Origins of Experimental Science, 1100–1700*, pp. 62–3
p. 40 'the same cause,...', ibid., p. 85
p. 40 'Metaphysics of Light', ibid., p. 128
p. 40 'This part of optics,...', ibid., p. 119
p. 40 'These modes of celestial...', ibid., p. 97
p. 41 'flying machines can be...', Crombie, *Medieval and Early Modern Science*, vol. 1, p. 55
p. 41 'science of weights', Grant, 'Jordanus de Nemore', *DSB*, 7, 172
p. 41 'positional gravity', ibid.
p. 41 'weight is heavier...', ibid.
p. 42 'would endure forever...', Moody, 'Jean Buridan', *DSB*, 2, 606
p. 42 'quantity of matter', ibid.
p. 43 'indisputably true that...', ibid., 2, 607
p. 44 'it is not impossible...', Clagett, 'Nicole Oresme,' *DSB*, 10, 223
p. 44 'subject to correction,...', Crombie, *Medieval and Early Modern Science*, vol. II, p. 78
p. 45 'For God fixed the...', ibid.
p. 45 'Completed in camp,...', Grant, 'Peter Peregrinus', *DSB*, 10, 533
p. 47 'It would be futile...', Crombie, *Robert Grosseteste and the Origins of Experimental Science, 1100–1700*, p. 216
p. 47 'multiplication of species' ibid.
p. 47 'there is something wonderful...', Minio-Paluello, 'William of Moerbeke', *DSB*, 9, 435
p. 47 'a globe of water...', Wallace, 'Dietrich of Freiberg', *DSB*, 4, 93
p. 49 'There is in Krakow...', Repcheck, p. 41
p. 49 'Nicolaus Nicolai de...', Gassendi and Thill, pp. 26–7
p. 51 'Ptolemy in his *Cosmography*...', Copernicus, *De revolutionibus*, I, 10
p. 52 'He once enjoyed...', Rosen, *Copernicus and the Scientific Revolution*, p. 141

CHAPTER 4. RENAISSANCE ITALY

p. 54 'he cared only...', quoted by Freely, *Jem Sultan*, p. 284
p. 54 'all the liberal arts', Shank, 'The Classical Scientific Tradition in Fifteenth-Century Vienna', in Ragep, F. Jamil, and Sally P. Ragep with Steven Livesey (eds) *Tradition, Transmission, Transformation*, p. 127
p. 55 'a Euclid and...', ibid.
p. 57 'the rebirth of trigonometry', Boyer, pp. 301–2

p. 58 'We decree and ordain...', Rosen, *Copernicus and His Successors*, p. 127
p. 58 *'Dominus Nicolaus Kopperlingk...'*, Gassendi and Thill, p. 37
p. 58 'The contemporary value...', Rosen, *Copernicus and His Successors*, pp. 127–8
p. 58 'Once matriculation had occurred...', ibid., p. 66
p. 58 'My teacher made observations...', Rosen, *Three Copernican Treatises*, p. 111
p. 59 'We made this observation...', Copernicus, *De revolutionibus*, IV, 27
p. 59 'But the elevations of...', ibid., II, 6
p. 60 'in order that you...', Rosen, *Copernicus and His Successors*, p. 20
p. 60 'Copernicus' well-worn copy...', ibid., pp. 21–2
p. 60 'in order to determine the moon...', Copernicus, *De revolutionibus*, IV, 14
p. 60 'at Rome, where,...', Rosen, *Three Copernican Treatises*, p. 111
p. 61 'The Pope is now 70...', Pastor, *History of the Popes*, vol. VI, p. 80
p. 61 'As a helpful physician...', Rosen, 'Nicholas Copernicus', *DSB*, 3, 402
p. 62 'If such an account...', Rosen, *Three Copernican Treatises*, p. 127
p. 62 'it would not be easy...', Copernicus, *De revolutionibus*, I, 10

CHAPTER 5. THE BISHOPRIC OF WARMIA

p. 67 'You will behold...', Copernicus, *Minor Works*, p. 50
p. 67 'the gay with...', ibid., p. 29
p. 67 'Ethical. From CRITIAS...', ibid., p. 30
p. 67 'This slender volume...', Rosen, *Copernicus and His Successors*, p. 25
p. 68 'All this with reference...', Copernicus, *De revolutionibus*, IV, 7
p. 68 'the number of observations...', Swerdlow and Neugebauer, I, p. 65
p. 69 'On the east is...', Gassendi and Thill, pp. 49–50
p. 70 'it was never greater...', Copernicus, *De revolutionibus*, II, 2
p. 71 'the difference [in his...', ibid., IV, 16
p. 71 'consulted the greatest...', Rosen, *Copernicus and His Successors*, p. 195
p. 71 'of all the theologians...', ibid.
p. 72 'I have great need...', ibid., p. 196
p. 72 'For not many years...', Copernicus, *De revolutionibus*, Preface
p. 72 'the ten or more...', ibid., III, 16
p. 72 'We find that...', ibid., III, 14
p. 73 'a manuscript of six...', Copernicus, *Minor Works*, p. 75
p. 73 'to the meridian...', Rosen, 'Nicholas Copernicus', *DSB*, 3, 402
p. 73 'as is clear from...', ibid.
p. 74 'A certain little treatise...', Rosen, *Copernicus and His Successors*, pp. 72–3
p. 74 'its date of composition...', Rosen, *Three Copernican Treatises*, p. 67
p. 74 'the sun is the center...', ibid., p. 58

CHAPTER 6. THE LITTLE COMMENTARY

p. 75 'For these theories...', Rosen, *Three Copernican Treatises*, p. 57
p. 76 'Having become aware...', ibid., pp. 57–8

p. 76 'the center of the universe...', ibid., p. 58
p. 77 'Philolaus believed in...', Gingerich, *The Eye of Heaven*, p. 186
p. 77 'There is no question...', ibid., p. 190
p. 77 'Whatever motion appears...', Rosen, *Three Copernican Treatises*, p. 59
p. 77 'I shall endeavor...', ibid.
p. 78 'From this reservation...', ibid.
p. 78 'The celestial spheres...', ibid., pp. 59–60
p. 78 'First it revolves annually...', ibid., p. 61
p. 79 'The second motion,...', ibid., p. 63
p. 80 'Saturn, Jupiter and Mars have...', ibid., p. 74
p. 80 'the radius of the first epicycle...', ibid., p. 77
p. 80 'There is a second...', ibid., pp. 77–8
p. 81 'In all cases...', ibid., p. 78
p. 81 'In latitude they have...', ibid.
p. 81 'The motion of the earth...', ibid., p. 80
p. 82 'Venus seems at times...', ibid., p. 83
p. 82 'Its latitude also changes...', Rosen, ibid., pp. 83–4
p. 83 'Of all the orbits...', ibid., p. 85
p. 83 'stationed about the center...', ibid., p. 87
p. 83 'Its motion in latitude...', ibid., p. 89
p. 83 'Thus Mercury runs...', ibid., p. 90

CHAPTER 7. THE LETTER AGAINST WERNER

p. 85 'Administrator of the...', Copernicus, *Minor Works*, p. 224
p. 85 'For the past seven...', Gingerich and MacLachlan, p. 85
p. 86 'canons want to act...', ibid., p. 90
p. 87 'elected administrator shall...', Copernicus, *Minor Works*, p. 224
p. 87 'Merten Caseler took possession...', ibid., p. 228
p. 87 'Bulls of [Pope] Innocent VI...'., ibid., p. 261
p. 88 'bad money drives...', Gingerich and MacLachlan, p. 89
p. 88 'For it undermines...', Copernicus, *Minor Works*, p. 17
p. 88 'But maybe someone...', ibid., p. 191
p. 89 'if possible, only one...', Gassendi and Thill, p. 85
p. 89 'if however, this could not...', Copernicus, *Minor Works*, p. 192
p. 91 'Some time ago...', Rosen, *Three Copernican Treatises*, p. 93
p. 91 'Consequently, lest I seem...', ibid, pp. 93–4
p. 91 'in the 150th year...', ibid., p. 94
p. 91 'The hypothesis in which...', ibid., p. 98
p. 92 'Therefore it is clear that the fixed...', ibid.
p. 92 'Hence it is clear...', ibid., p. 101
p. 92 'Now we said that...', Copernicus, *De revolutionibus*, I, 11
p. 93 'Moreover, there is...', ibid., III, 1

Source Notes

p. 93 'For the sake of...', ibid.
p. 93 'The apparent or unequal...', Rosen, *Three Copernican Treatises*, p. 102
p. 93 'Therefore it is clear that the first...', ibid., p. 102
p. 93 'Already an eleventh sphere...', Copernicus, *De revolutionibus*, III, 1
p. 95 'Hence he saw...', ibid.
p. 95 'But Copernicus was...', Rosen, *Copernicus and His Successors*, p. 105
p. 95 'Here', he explains,...', ibid.
p. 95 'Now Copernicus replied...', ibid., p. 110
p. 96 'that, according to...', ibid.
p. 96 'according to a more...', ibid.
p. 96 'more than 49...', ibid.
p. 96 'The higher value...', ibid., pp. 110–1
p. 96 'So much for the...', ibid., pp. 105–6

CHAPTER 8. THE FRAUENBURG WENCHES

p. 99 'just price', Copernicus, *Minor Works*, p. 283
p. 100 'From one sack...', ibid., p. 281
p. 100 'Doctor Nicolaus and other...', Gassendi and Thill, p. 66
p. 100 'Besides, you should talk...', ibid., p. 77
p. 101 'the only one outstanding...', Rosen, *Copernicus and His Successors*, p. 158
p. 101 'Nicholas Copernicus, canon...', ibid.
p. 103 'My lord, Most Reverend...', Copernicus, *Minor Works*, pp. 319–20
p. 103 'With due expression...', ibid.
p. 104 'Doctor Nicolaus and Doctor Wille...', Gassendi and Thill, p. 66
p. 104 'I am sending you...', ibid., pp. 124–5
p. 105 'Some years ago,...', Rosen, *Copernicus and the Scientific Revolution*, pp. 187–8
p. 106 'First among them...', Copernicus, *De revolutionibus*, Preface
p. 106 '*ein alter thumherr*...', Gassendi and Thill, p. 172
p. 106 'The Chapter of Warmia...', Biskup, *Regesta Copernicana*, p. 162
p. 107 'whose designation would...', Gassendi and Thill, p. 102
p. 108 '*gutten Gesellen*', ibid., p. 98
p. 109 'I have received...', Copernicus, *Minor Works*, pp. 323–4
p. 109 'a few more days,...', Gassendi and Thill, p. 69
p. 109 'I feel better,....', ibid.
p. 109 'cherish [Copernicus]...as a brother', Rosen, *Copernicus and the Scientific Revolution*, p. 168
p. 110 'My lord, Most Reverend...', ibid., p. 151
p. 110 'I have done what...', Copernicus, *Minor Works*, pp. 334–5
p. 111 'He will be overcome...', Rosen, *Copernicus and the Scientific Revolution*, p. 152
p. 111 'God Almighty will strengthen...', ibid., pp. 152–3
p. 111 'I am sending back...', ibid., p. 154
p. 112 'As regards the Frauenburg...', ibid., pp. 156–7

p. 112 'He [Copernicus] is renowned…', ibid., pp. 158–9
p. 112 'I have talked earnestly…', ibid., p. 160

CHAPTER 9. THE FIRST DISCIPLE

p. 115 'Finally, hearing the great fame…', Danielson, *The First Copernican*, p. 98
p. 117 'born to study mathematics', ibid., p. 24
p. 117 'Greetings! This youth is…', Repcheck, pp. 120–1
p. 118 'to the perfecting of…', Danielson, *The First Copernican*, p. 34
p. 118 'No one can bypass…', ibid.
p. 118 'When I was with you…', Rosen, *Three Copernican Treatises*, p. 162
p. 119 'The essential feature…', Evans, p. 403
p. 119 'I have chosen Nuremberg…', Rosen, *Copernicus and the Scientific Revolution*, p. 171
p. 120 'he had made 746…', Pannekoek, p. 181
p. 120 'The ancients have…', Copernicus, *De revolutionibus*, V, 30
p. 120 'The first observation was…', ibid.
p. 121 'The second was taken…', ibid.
p. 121 'Copernicus relied almost…', Gingerich, *The Eye of Heaven*, p. 185
p. 122 'The Book was still…', ibid., p. 16
p. 122 'The list of 22 items,…', Pannekoek, p. 180
p. 123 'taking the center…', Saliba, *A History of Arabic Astronomy*, p. 75
p. 125 'On May 14th I wrote…', Rosen, *Three Copernican Treatises*, p. 109
p. 125 'Your letter full of…', Gassendi and Thill, p. 183
p. 126 'One of them is…', Rosen, *Three Copernican Treatises*, p. 192
p. 126 'In addition, the benevolent…', ibid., p. 195
p. 126 'the esteemed and energetic…', ibid.
p. 126 'Homer's Achilles, as it were', ibid.
p. 127 'In the proceeding against…', Rosen, *Copernicus and the Scientific Revolution*, p. 161
p. 127 'He realized that it…', Rosen, *Three Copernican Treatises*, p. 192
p. 127 'Since my teacher was social…', ibid.
p. 128 'He therefore decided…', ibid., pp. 192–3
p. 128 'Then His Reverence pointed out…', ibid., p. 193
p. 129 'By these and many other…', ibid., p. 195
p. 129 'TO THE MOST ILLUSTRIOUS…', Danielson, *The First Copernican*, p. 70
p. 129 'The opinions of older…', Rosen, *Three Copernican Treatises*, p. 196
p. 129 'From my library…', ibid.

CHAPTER 10. THE FIRST ACCOUNT

p. 131 'But now this…', Rosen, *Copernicus and the Scientific Revolution*, p. 192
p. 132 'Greetings, most excellent…', ibid., pp. 192–3

Source Notes

p. 132 'I have always felt...', ibid., p. 193
p. 132 'The peripatetics and theologians...', ibid., p. 194
p. 133 'Now a year has passed...', Swerdlow, N. M., 'Annals of Scientific Publishing: Johan Petreius's Letter to Rheticus,' *Isis*, Vol. 83, No. 2 (June 1992), pp. 273–4
p. 133 'Since the astronomical speculations...', Danielson, *The First Copernican*, p. 287
p. 134 'And so, if that...', ibid., p. 117
p. 134 'very precise calculation', ibid.
p. 134 'I do not argue about...', ibid.
p. 134 'The book certainly departs...', ibid., p. 212
p. 135 'to all who cherish...', ibid., p. 213
p. 135 'First of all...', Rosen, *Three Copernican Treatises*, p. 109
p. 135 'My teacher has written...', ibid., pp. 109–10
p. 135 'contains the general...', ibid., p. 110
p. 136 'the doctrine of...', ibid.
p. 136 'depends, in part,...', ibid.
p. 136 'mastered the first three...', ibid.
p. 136 'partly because my...', ibid.
p. 136 'Therefore,' he says, 'I shall...', ibid., p. 111
p. 136 'Realizing that equality...', ibid., pp. 116–17
p. 136 'that the entire period...', ibid., p. 117
p. 137 'Since every difficulty...', ibid., p. 119
p. 137 'My teacher further...', ibid., p. 121
p. 137 'I shall add a...', ibid., pp. 121–2
p. 138 'It was a difficult...', ibid., p. 131
p. 138 'He supposes therefore...', ibid., p. 134
p. 138 'The second inequality...', ibid., pp. 134–5
p. 139 'Furthermore, most learned...', ibid., p. 135
p. 139 'The planets are each...', ibid., pp. 135–6
p. 140 'In the first place,...', ibid., p. 136
p. 140 'Thirdly, the planets...', ibid., pp. 136–7
p. 140 'I shall proceed...', ibid., p. 142
p. 141 'established by hypothesis...', ibid., p. 143
p. 141 'Hence this sphere...', ibid.
p. 141 'Then, in harmony with...', ibid.
p. 141 'The other spheres...', ibid., pp. 143–4
p. 141 'For if we follow...', ibid., p. 144
p. 142 'remarkable symmetry and...', ibid., p. 145
p. 142 'However, in the hypotheses...', ibid., pp. 146–7
p. 143 'With regard to the apparent...', ibid., pp. 164–5
p. 143 'axiom that all...', ibid., p. 166
p. 143 'Just as my teacher...', ibid., p. 168
p. 144 'in a lust of novelty...', ibid., p. 187

p. 144 'Such is his time...', ibid.
p. 144 'You might say that...', ibid., p. 190
p. 145 'But if it is to be...', ibid., pp. 194–5
p. 145 'It is altogether likely...', ibid., p. 10

CHAPTER 11. PREPARING THE REVOLUTIONS

p. 147 'God-given skill...', Copernicus, *Minor Works*, pp. 343–8
p. 148 'is the first detailed...', Swerdlow and Neugebauer, I, p. 28
p. 149 'Nicolaus Copernicus, the...', Danielson, *The First Copernican*, p. 220
p. 149 'interpretation of Ptolemy', ibid., p. 92
p. 149 'Let us rouse...', ibid., p. 93
p. 149 'a modern astronomer...', Gingerich, *The Eye of Heaven*, p. 222
p. 150 'The science of triangles,...', Danielson, *The First Copernican*, p. 37
p. 150 'the most illustrious...', ibid., p. 95
p. 150 'Such a learned...', ibid.
p. 150 'Copernicus's first publication...', Swerdlow and Neugebauer, pp. 27–8
p. 150 'Here trigonometry really...', Boyer, p. 321
p. 151 'computations produced tables...', Danielson, *The First Copernican*, p. 201
p. 151 'Hospes: But this Rheticus...', ibid., p. 143
p. 151 'I searched for someone...', ibid., p. 138
p. 152 'There are those...', ibid., p. 139
p. 152 'I remember myself...', Repcheck, pp. 1–43
p. 152 'He wanted his researches...', ibid., p. 144
p. 153 'But when I see...', Rosen, *Three Copernican Treatises*, p. 163
p. 153 'In Wittenberg, the...', Rosen, *Copernicus and the Scientific Revolution*, pp. 118–19
p. 154 'fool', ibid., p. 119
p. 154 'For every apparent...', Copernicus, *De revolutionibus*, I, 5
p. 154 'There was mention...', Rosen, *Copernicus and the Scientific Revolution*, pp. 182–3
p. 155 'Certain people think...', ibid., p. 120
p. 155 'On folio 2 verso...', ibid., p. 184
p. 156 'lured away from...', Danielson, *The First Copernican*, p. 104
p. 156 'To the Reader...', Rosen, *Copernicus and the Scientific Revolution*, p. 195
p. 157 'For this art...', ibid., pp. 195–6
p. 158 'You have in this...', Gingerich, *The Book Nobody Read*, p. 20
p. 158 'Concerning this letter...', Danielson, *The First Copernican*, p. 113
p. 159 'I was shocked...', Rosen, *Copernicus and the Scientific Revolution*, pp. 165–6
p. 159 'On my return from...', ibid., p. 167
p. 160 'I should like...', ibid., pp. 167–8
p. 161 'It is not unknown...', ibid., pp. 168–9
p. 161 'She, who has been...', ibid., p. 169

Source Notes

CHAPTER 12. THE REVOLUTIONS OF THE CELESTIAL SPHERES

p. 163 'I can reckon easily...', Copernicus, *De revolutionibus*, Preface, p. 2
p. 164 'knowledge that mathematicians...', ibid., Preface, p. 3
p. 164 'reread all the books...', ibid., Preface, p. 4
p. 164 'Therefore, I also,...', ibid., Preface, pp. 4–5
p. 165 'And so having laid...', ibid., Preface, p. 5
p. 165 'Mathematics is written for...', ibid., Preface, pp. 5–6
p. 166 'reasons of geometry...', ibid., I, 3
p. 166 'it is manifest...', ibid.
p. 166 'The Movement of the Celestial...', ibid., I, 4
p. 166 'Although there are so...', *De revolutionibus*, I, 5
p. 167 'made the Earth to revolve...', ibid.
p. 167 'Now upon this assumption...', ibid.
p. 167 'It can be understood...', ibid., I, 6
p. 168 'I myself think...', ibid., I, 9
p. 168 'In the center...', ibid., I, 10
p. 169 'made no distinction...', ibid., III, 1
p. 169 'Hipparchus[...]was the first...', ibid.
p. 170 'For[...]the two revolutions...', ibid.
p. 170 'correction of the natural...', Swerdlow and Neugebauer, I, p. 172
p. 171 'the mean position...', Copernicus, *De revolutionibus*, IV, 10
p. 172 'In our explanation...', ibid., IV, Introduction
p. 172 'The planetary models...', Swerdlow and Neugebauer, I, p. 47
p. 173 'How Copernicus learned...', Swerdlow, 'Copernicus, Nicolaus (1473–1543),' in *Encyclopedia of the Scientific Revolution*, p. 165
p. 173 'Copernicus gives a total...', Swerdlow and Neugebauer, I, p. 200
p. 174 'is without doubt...', ibid., I, p. 232
p. 174 'The problem of the occultation...', ibid.
p. 174 'In the case of the moon...', Copernicus, *De revolutionibus*, IV, 30
p. 175 'the time of true conjunction...', ibid.
p. 175 'Let all of this...', ibid., IV, 32
p. 175 'Up to now...', ibid., V, Introduction
p. 176 'Two longitudinal movements...', ibid., V, 1
p. 176 'Furthermore there is a different...', ibid.
p. 178 'These and similar...', ibid., V, 2
p. 178 'The period of P...', Swerdlow and Neugebauer, I, p. 291
p. 179 'the center of the epicycle...', ibid., I, p. 292
p. 179 'the distance SP...', ibid.
p. 179 'For Ptolemy's model:...', ibid., I, p. 297
p. 180 'We have indicated...', Copernicus, *De revolutionibus*, VI, Introduction
p. 181 'Copernicus, ignorant of...', Swerdlow and Neugebauer, I, p. 483

p. 181 'specifically to the...', ibid.
p. 181 'was the result...', ibid
p. 181 'But Copernicus's representation...', ibid
p. 182 'The latitude theory...', ibid., I, p. 535
p. 183 'Book VI must have...', ibid., I, p. 537

CHAPTER 13. THE COPERNICAN REVOLUTION

p. 185 'I have not wanted...', Danielson, *The First Copernican*, p. 139
p. 185 'the most learned man...', Gingerich, *The Eye of Heaven*, p. 210
p. 185 'the computation is not...', ibid.
p. 185 'Although it is clear...', ibid., p. 212
p. 186 'All posterity will...', ibid., p. 221
p. 186 'Very convenient tables...', ibid., p. 223
p. 186 'Copernicus [...] affirmeth that...', Kuhn, *The Copernican Revolution*, p. 186
p. 186 'That is trulye to be gathered...', Francis R. Johnson, 'The influence of Thomas Digges in the progress of modern Astronomy in Sixteenth-Century England', in *Osiris* 1, 390–410 (June 1936)
p. 187 'Wherefore, I have published...', Gingerich, *The Eye of Heaven*, p. 230
p. 187 'more than Herculean', Armitage, *Copernicus and Modern Astronomy*, p. 165
p. 187 '*Vulgi opinio error*', Gingerich, *The Book Nobody Read*, p. 119
p. 187 'so that Englishmen...', ibid.
p. 187 'In this our age,...', Dreyer, *A History of Astronomy from Thales to Kepler*, p. 347
p. 188 'There are then innumerable...', Koyré, *From the Closed World to the Infinite Universe*, p. 49
p. 188 'theory of Aristarchus,...', Dreyer, *A History of Astronomy from Thales to Kepler*, p. 350
p. 189 'something divine that...', Gingerich, *The Eye of Heaven*, p. 172
p. 189 'a second Ptolemy', Dreyer, *Tycho Brahe*, p. 74
p. 190 'I knew perfectly well...', Ferguson, pp. 46–7
p. 190 'I doubted no longer...', ibid., p. 47
p. 191 'exactly circular but...', Dreyer, *A History of Astronomy from Thales to Kepler*, p. 366
p. 191 'There really are not...', Rosen, 'Nicolaus Copernicus', *DSB*, 3, 409
p. 194 'How many mathematicians...', A. R. Hall, p. 126
p. 194 'The earth's orbit...', Gingerich, 'Johannes Kepler', *DSB*, 7, 290
p. 195 'a still unexhausted...', Caspar, p. 64
p. 195 'When I was studying...', Gingerich, 'Johannes Kepler', *DSB*, 7, 290
p. 195 'no one will wonder...', ibid., 7, 291
p. 195 'Have faith, Galilei...', Koestler, p. 364
p. 196 'a brilliant speculation', Ferguson, p. 255
p. 196 'You will come...', Gingerich, 'Johannes Kepler', *DSB*, 7, 293
p. 196 'Tycho Brahe, himself,...', Gingerich, *The Eye of Heaven*, p. 310

Source Notes

p. 197 'although he knew...', Ferguson, p. 284
p. 197 'Divine Providence granted...', Gingerich, 'Johannes Kepler', *DSB*, 7, 295
p. 198 'I lack only...', ibid.
p. 198 'I was almost driven...', ibid., 7, 297
p. 199 'Hereafter, when they...', Milton, *Paradise Lost*, 79–84
p. 199 'a book full of astronomical...', Gingerich, 'Johannes Kepler', *DSB*, 7, 297
p. 200 'I thank you...', ibid., 7, 299
p. 201 'necessary for the contemplation...', Caspar, pp. 276–7
p. 201 'those opinions of yours...', John Donne, 'Ignatius His Conclave', in *Complete Poetry and Selected Prose of John Donne*, p. 365
p. 201 'And new Philosophy...', 'The Anatomy of the World', in *The Complete Poetry of John Donne*, pp. 277–8
p. 202 'the novelty of my discoveries...', Gingerich, *The Eye of Heaven*, p. 321
p. 202 'I used to measure...', Gingerich, 'Johannes Kepler', *DSB*, 7, 307
p. 203 'to the bodies of the sun...', Dreyer, *A History of Astronomy from Thales to Kepler*, pp. 399–400
p. 203 'a mathematical demonstration...', Gingerich, 'Johannes Kepler', *DSB*, 7, 308
p. 203 'found to agree...', ibid.

CHAPTER 14. DEBATING THE COPERNICAN AND PTOLEMAIC MODELS

p. 207 'In which all...', Machamer, 'Galileo's machines...', in Machamer (ed.) *The Cambridge Companion to Galileo*, p. 58
p. 208 'Thus, moment [*momento*] is...', ibid., p. 61
p. 208 'Many years ago...', Pannekoek, pp. 225–6
p. 208 'geometric and military...', Stillman Drake (tran.), *Discoveries and Opinions of Galileo*, p. 16
p. 209 'THE STARRY MESSENGER:....', ibid., p. 21
p. 210 'Behold then, four stars...', ibid., p. 24
p. 210 'On the fourth...', ibid., p. 32
p. 210 'the surface of the moon...', ibid., p. 31
p. 210 'Deserving of notice...', ibid., p. 47
p. 212 'a host of other...', ibid.
p. 212 'was overwhelmed by...', ibid.
p. 212 'Hence to the three...', ibid., pp. 47–8
p. 212 'With the aid...', ibid., p. 49
p. 213 'Here we have...', ibid., p. 57
p. 214 'Things so hard...', Pannekoek, p. 230
p. 215 'about the time...', Stillman Drake (tran.), *Discoveries and Opinions of Galileo*, pp. 75–6

p. 215 'Using the concept...', Stillman Drake, 'Galileo Galilei', *DSB*, 5, 41
p. 216 'I say, then,...', Stillman Drake (tran.), *Discoveries and Opinions of Galileo*, p. 144
p. 216 'For I seem...', ibid., p. 113
p. 217 'thus brought the question...', ibid., p. 85
p. 217 'I have been told...', ibid., p. 146
p. 218 'was not written...', ibid., p. 212
p. 219 'In Sarsi I seem...', ibid., p. 237
p. 219 'the least important...', ibid.
p. 219 'Well, Sarsi, that...', ibid., pp. 237–8
p. 220 'Let us call...', Crombie, *Medieval and Early Modern Science*, vol. II, p. 145
p. 220 'You are wrong,...', ibid., vol. II, p. 136
p. 221 'Nor can I...', ibid., vol. II, p. 141
p. 221 'until it might...', Machamer, 'Introduction', in Machamer (ed.) *The Cambridge Companion to Galileo*, p. 23
p. 221 'it would still be...', Galileo, *Dialogue Concerning the Two Chief World Systems*, p. 464
p. 222 'Your Galileo has ventured...', De Santillana, *The Crime of Galileo*, p. 191
p. 222 'I, Galileo, son of...', Machamer, 'Introduction', in Machamer (ed.) *The Cambridge Companion to Galileo*, p. 23
p. 223 'when starting from...', Crombie, *Medieval and Early Modern Science*, vol. II, p. 145
p. 223 'only those means...', ibid.
p. 223 'At present it is...', ibid., vol. II, pp. 146–8
p. 224 'The knowledge of...', ibid., vol. II, p. 156
p. 224 'had altogether given...', Koestler, p. 503
p. 224 'Take note theologians...', Galileo, *Dialogue Concerning the Two Chief World Systems*, p. v

CHAPTER 15. THE NEWTONIAN SYNTHESIS

p. 225 'There is no effect...', Gaukroger, *The Emergence of a Scientific Culture*, p. 253
p. 225 'all properties arise...', A. R. Hall, p. 207
p. 226 'The subtlety of nature...', ibid., p. 34
p. 226 'hypothesis', Mary Hesse, 'Francis Bacon', *DSB*, 1, 372
p. 226 ' 'calculations and predictions'...', ibid.
p. 226 'In physics I...', Crombie, 'Descartes', *DSB*, 4, 53
p. 226 'quantity of motion', Descartes, *The Philosophical Writings of Descartes*, vol. 1, p. 240
p. 227 'In the beginning,...', ibid.
p. 227 'Each and every...', ibid.
p. 227 'all motion is...', ibid., vol. 1, pp. 241–2
p. 227 'if a body...', ibid., vol. 1, p. 242

Source Notes

p. 227 'true mathematics', Michael S. Mahoney, 'Descartes: Mathematics and Physics', *DSB*, 4, 55

p. 228 'new world', Descartes, *The Philosophical Writings of Descartes*, vol. 1, p. 90

p. 228 'if we consider…', Gaukroger, *The Emergence of a Scientific Culture*, p. 305

p. 229 'almost nothing I can…', Gaukroger, *Descartes, an intellectual biography*, p. 421

p. 229 'the simple pendulum…', Gaukroger, *The Emergence of a Scientific Culture*, p. 439

p. 230 'Otherwise[…] let them…', Clarke, Desmond M., 'Pascal's Philosophy of Science', in *The Cambridge Companion to Pascal*, ed. Nicholas Hammond, p. 114

p. 230 'Air ether consists…', Gaukroger, *The Emergence of a Scientific Culture*, p. 370

p. 231 'By this very thing…', Crombie, *Medieval and Early Modern Science*, vol. II, p. 299

p. 232 'In the beginning…', Westfall, p. 143

p. 233 'rays which make…', ibid., p. 160

p. 233 'A naturalist would…', Gaukroger, *The Emergence of a Scientific Culture*, pp. 393–4

p. 234 'For the best…', Westfall, p. 242

p. 234 'if it were true…', Gaukroger, *The Emergence of a Scientific Culture*, p. 380

p. 235 'extremely well pleased…', Manuel, p. 144

p. 235 'What Descartes did…', Westfall, p. 274

p. 235 'particularly if you will…', Gaukroger, *The Emergence of a Scientific Culture*, p. 434

p. 235 'your Hypotheses of…', ibid., p. 435

p. 235 'the inflection of a direct…', ibid.

p. 236 'This depends upon three…', Westfall, p. 382

p. 237 '[O]ur present work…', Newton, *Mathematical Principles of Natural Philosophy*, p. 382

p. 237 'quantity of matter', ibid., p. 403

p. 237 'quantity of motion', ibid., p. 404

p. 237 'inherent force of matter…', ibid.

p. 238 'Impressed force is the…', ibid., p. 405

p. 238 'are impelled, or…', ibid.

p. 238 'Law 1: Every body perseveres…', ibid., pp. 416–17

p. 239 'For attractions are…', ibid., p. 561

p. 239 'by radii drawn…', ibid., p. 797

p. 240 'Gravity exists in all…', ibid., p. 810

p. 240 'The planets move…', ibid., p. 817

p. 240 'The axes of the planets…', ibid., p. 821

p. 240 'the ebb and flow…', ibid., p. 835

p. 240 'To find the precession…', ibid., p. 885

p. 240 'the precession of the…', ibid., p. 887

p. 240 'The comets are higher…', ibid., p. 888

p. 240 'This most elegant…', ibid., p. 940

p. 241 'corrected by the…', Newton, *Opticks*, p. lxxvii

p. 241 'My design in...', ibid., p. 1
p. 241 'the oldest and...', ibid., p. 369
p. 241 'All these things...', ibid., p. 400
p. 242 'Light is propagated...', ibid., p. 277
p. 242 'method of fluxions', Westfall, p. 143
p. 243 'four metals' of alchemy...', A. P. Youschevitz, 'Isaac Newton, *DSB*, 10, 81
p. 243 'Let Mortals rejoice...', Westfall, p. 874
p. 243 'If I have seen...', ibid., p. 274

BIBLIOGRAPHY

Africa, Thomas W., 'Copernicus' Relation to Aristarchus and Pythagoras', *Isis* 52, No. 3 (Sept. 1961), pp. 403–9.
Aristotle, *Complete Works*, 2 vols., edited by Jonathan Barnes (Princeton, 1984).
Armitage, Angus, *Sun Stand Thou Still; The Life and Works of Copernicus the Astronomer* (New York, 1947).
———, *Copernicus and Modern Astronomy* (New York, 2004).
Baker, Robert H., *Astronomy, An Introduction* (New York, 1930).
Bald, R. C., *John Donne, A Life* (Oxford, 1971).
Biskup, Marian, *Regesta Copernicana* (Calendar of Copernicus' Papers) (Wroclaw, Poland, 1973).
Bos, H. J. M., 'Christiaan Huygens', *DSB*, 6, pp. 597–612.
Boyer, Carl B., *A History of Mathematics* (New York, 1968).
Burmeister, K. H., 'Georg Joachim Rheticus', *DSB*, 11, pp. 395–8.
Burnet, John, *Greek Philosophy, Thales to Plato* (London, 1981).
Butterfield, Herbert, *The Origins of Modern Science, 1300–1800* (New York, 1957).
Callus, Daniel A. (ed.), Robert Grosseteste, *Scholar and Bishop* (Oxford, 1955).
Cantor, Norman F., *The Civilization of the Middle Ages* (New York, 1993).
Carmody, Francis J., *The Astronomical Works of Thabit ibn Qurra* (Berkeley, 1960).
Caspar, Max, *Kepler*, translated by C. Doris Hellman (New York, 1962).
Chatillon, Jean, 'Giles of Rome', *DSB*, 5, 402–3.
Clagett, Marshall, *The Science of Mechanics in the Middle Ages* (Madison, Wisconsin, 1959).
——— (ed.), *Archimedes in the Middle Ages, Vol. I: The Arabo-Latin Tradition* (Madison, Wisconsin, 1964).
———, *Greek Science in Antiquity* (London, 2001).
———, 'Adelard of Bath', *DSB*, 1, 61–64.
———, 'John of Palermo', *DSB*, 7, 133–134.
———, 'Nicole Oresme', *DSB*, 10, 223–230.
Clarke, Desmond M., 'Pascal's Philosophy of Science', in *The Cambridge Companion to Pascal*, ed. Nicholas Hammond, p. 114.
Cohen, I. Bernard, *Revolution in Science* (Cambridge, Massachusetts, 1985).
Colish, Marcia L., *Medieval Foundations of the Western Intellectual Tradition 400–1400* (New Haven and London, 1997).

Copernicus, Nicolaus, *Minor Works*, translation and commentary by Edward Rosen with the assistance of Erna Hilfstein (Baltimore, Maryland, 1992).
———, *De revolutionibus (On the Revolutions of the Celestial Spheres)*, translated by Glen Wallis (Philadelphia, 2002).
Crombie, A. C., *Robert Grosseteste and the Origins of Experimental Science, 1100–1700* (Oxford, 1953).
———, *Medieval and Early Modern Science*, 2 vols., 2nd ed. (Cambridge, Massachusetts, 1963).
——— (ed.), *Scientific Change; historical sketches in the intellectual, social and technical conditions for scientific discovery and technical invention from antiquity to the present* (New York, 1963).
———, 'Descartes', *DSB*, 4, 51–5.
———, 'Robert Grosseteste', *DSB*, 5, 548–54.
Crombie, A. C. and J. D. North, 'Roger Bacon', *DSB*, 1, 377–85.
Dales, Richard, *The Scientific Achievements of the Middle Ages* (Philadelphia, 1973).
Daly, John F., 'Johannes de Sacrobosco (John of Hollywood)', *DSB*, 12, 60–3.
Danielson, Dennis, 'Achilles Gasser and the Birth of Copernicanism', *Journal of the History of Astronomy*, 35 (2004), pp. 457–74.
———, *The First Copernican: Georg Joachim Rheticus and the Rise of the Copernican Revolution* (New York, 2006).
Dante, *The Divine Comedy*, translated by H. F. Cary (New York, 1908).
De Santillana, Giorgio, *The Crime of Galileo*, Chicago, 1955 *Dictionary of Scientific Biography* (DSB), 16 vols., edited by Charle Coulston Gillespie (New York, 1970–80).
Dijksterhuis, F. J., *Archimedes* (Princeton, 1987).
Donne, John, *The Complete Poetry and Selected Prose of John Donne*, edited by John Hayward (London, 1929).
———, *The Complete Poetry of John Donne*, edited by John T. Shawcross (London, 1968).
Drake, Stillman (translator), *Discoveries and Opinions of Galileo* (Garden City, New York, 1952).
———, 'Giovanni Battista Benedetti', *DSB*, 1, 604–9.
———, 'Galileo Galilei', *DSB*, 5, 237–248.
Dreyer, J. L. E., *A History of Astronomy from Thales to Kepler* (New York, 1953).
———, *Tycho Brahe* (New York, 1963).
Easton, Joy B., 'John Dee', *DSB*, 4, 5–6.
———, 'Thomas Digges,' *DSB*, 4, 97–8.
———, 'Robert Recorde,' *DSB*, 11, 338–40.
Eicholz, David E., 'Pliny', *DSB*, 11, 38–40.
Encyclopedia of the Scientific Revolution, Wilbur Applebaum (ed.) (New York, 2000).
Evans, James, *The History and Practice of Ancient Astronomy* (New York and Oxford, 1999).
Ferguson, Kitty, *Tycho and Kepler: The Unlikely Partnership That Forever Changed Our Understanding of the Heavens* (New York, 2002).

Folkerts, Menso, 'Regiomontanus' Role in the Transmission and Transformation of Greek Mathematics', in Ragep, F. Jamil, and Sally P. Ragep with Steven Livesey (eds) *Tradition, Transmission, Transformation* (Leiden, 1996), pp. 89–113.
Fox, Robert, *Thomas Harriot: An Elizabethan Man of Science* (Farnham, Surrey, 2000).
Freely, John, *The Emergence of Modern Science, East and West* (Istanbul, 2004).
———, *Aladdin's Lamp: How Greek Science Came to Europe Through the Islamic World* (New York, 2009).
———, *Light from the East: How the Science of Medieval Islam helped to Shape the Western World* (London, 2010).
Furley, David J., 'Lucretius', *DSB*, 8, 536.
Gade, John A., *The Life and Times of Tycho Brahe* (Princeton, 1947).
Galilei, Galileo, *Discourses Concerning Two New Sciences, of Mechanics and of Motion* (New York, 1914).
———, *Dialogue Concerning the Two World Systems, Ptolemaic and Copernican*, translated by Stillman Drake (Berkeley, 1967).
Gassendi, Pierre and Olivier Thill, *The Life of Copernicus (1473–1543)* (Fairfax, Virginia, 2003).
Gaukroger, Stephen, *Descartes, An Intellectual Biography* (Oxford, 1995).
———, *The Emergence of a Scientific Culture: Science and the Shaping of Modernity 1210–1685* (Oxford, 2006).
Geanakoplos, Dino John, *Greek Scholars in Venice: Studies in the Dissemination of Greek Learning from Byzantium to Western Europe* (Cambridge, Massachusetts, 1962).
Geymont, Ludovico, *Galileo, a biography and inquiry into his philosophy of science* (New York, 1965).
Gingerich, Owen, *The Eye of Heaven: Ptolemy, Copernicus, Kepler* (New York, 1993).
———, *The Book Nobody Read: Chasing the Revolutions of Nicolaus Copernicus* (New York, 2004).
———, 'Johannes Kepler', *DSB*, 7, 289–312.
———, 'Erasmus Reinhold', *DSB*, 11, 365–7.
Glick, Thomas F., 'Leo the African', *DSB*, 8, 190–2.
Gliozzi, Marion, 'Evangelista Torricelli', *DSB*, 13, 433–40.
Grant, Edward, *Physical Science in the Middle Ages* (New York, 1971).
———, 'Jordanus de Nemore,' *DSB*, 7, 171–9.
———, 'Peter Peregrinus', *DSB*, 10, 532–40.
Guerlac, Henry, 'Copernicus and Aristotle's Cosmos', in *Journal of the History of Ideas*, 29, No. 1, 109–113 (1968).
Gutas, Dimitri, *Greek Thought, Arabic Culture: the Graeco-Arabic Translation Movement in Baghdad and Early Abbasid Society* (London, 1998).
Guthrie, William K. C., *A History of Greek Philosophy*, 6 vols. (Cambridge, Massachusetts, 1962–81).
Hall, A. R., The Scientific Revolution, 1500–1800 (Boston, 1956).
Hall, Marie Boas, 'Robert Boyle', *DSB*, 2, 377–82.
Hannam, James, *God's Philosophers: How the Medieval World Laid the Foundations of Modern Science* (London, 2009).

Haring, Nikolaus M., 'Thierry of Chartres', *DSB*, 13, 339–41.
Harkness, Deborah E., *The Jewel Box: Elizabethan London and the Scientific Revolution* (New Haven, 2007).
Haskins, Charles Homer, *The Rise of Universities* (New York, 1923).
———, *Studies in the History of Mediaeval Science* (Cambridge, 1924).
———, *The Renaissance of the Twelfth Century* (New York, 1957).
Heath, T. L., *Aristarchus of Samos, the ancient Copernicus* (Oxford, 1959).
Hellman, C. Doris. 'Tycho Brahe', *DSB*, 2, 401–14.
Hellman, C. Doris and Noel M. Swerdlow, 'Georg Peurbach', *DSB*, 15, 473–9.
Heninger, S. K., *Touches of Sweet Harmony: Pythagorean Cosmology and Renaissance Poetics* (San Marino, California, 1974).
Henry, John, *The Scientific Revolution and the Origins of Modern Science* (New York, 2002).
Hesse, Mary, 'Francis Bacon', *DSB*, 1, 372–7.
Hofmann, Joseph E., 'Nicholas Cusa', *DSB*, 3, 512–16.
———, 'Gottfried Wilhelm Leibnitz', *DSB*, 8, 149–68.
Huff, Toby E., *The Rise of Early Modern Science; Islam, China and the West* (Cambridge, 1993).
Jayawardene, S., 'Luca Pacioli', *DSB*, 10, 269–72.
Jeanneu, Edouard, 'Bernard Silvestre', *DSB*, 2, 21–2.
Johnson, Francis R., 'The influence of Thomas Digges in the progress of modern Astronomy in Sixteenth-Century England', in *Osiris* 1, 390–410 (June 1936).
———, *Astronomical Thought in Renaissance England* (Baltimore, 1937).
———, 'Marlowe's Astronomy and Renaissance Skepticism', in *Journal of English Literary History*, Vol. 3, No. 4 (Dec. 1946), 241–54.
Jones, Charles W., 'The Venerable Bede', *DSB*, 1, 564–6.
Kantorowicz, Ernst, *Frederick the Second, 1194–1250*, translated by E. O. Lorimer (New York, 1931).
Keele, Kenneth D., Ladidlao Reti, Augusto Marinoni, and Marshall Claggett, 'Leonardo da Vinci', *DSB*, 8, 192–245.
Kelly, Suzanne. 'William Gilbert', *DSB*, 5, 396–401.
Kirk, G. S. and J. E. Raven, *The Presocratic Philosophers* (Cambridge, 1962).
Kline, Morris, *Mathematical Thought from Ancient to Modern Times*, 3 vols. (New York, 1990).
Koestler, Arthur, *The Sleepwalkers: A History of Man's Changing Vision of the Universe* (London, 1959).
Kopal, Zdenek, 'Olaus Roemer', *DSB*, 11, 525–7.
Koyré, Alexandre, *From the Closed World to the Infinite Universe* (New York, 1958).
———, *Newtonian Studies* (Cambridge, Massachusetts, 1965).
———, *The Astronomical Revolution: Copernicus, Kepler, Borelli*, translated by R. E. W. Madison (Ithaca, New York, 1973).
Krafft, Fritz, 'Otto von Guericke', *DSB*, 5, 574–6.
Kren, Claudia, 'Alain de Lille', *DSB*, 1, 91–2.

Bibliography

———, 'Dominicus Gundissalinus', *DSB*, 5, 591–3.
———, 'Hermann the Lame', *DSB*, 6, 301–3.
———, 'Roger of Hereford', *DSB*, 11, 503–4.
Kuhn, Thomas S., *The Copernican Revolution: Planetary Astronomy in the Development of Western Thought* (Cambridge, Massachusetts, 1957).
———, *The Structure of Scientific Revolutions* (Chicago, 1976).
Leff, Gordon, 'John Duns Scotus', *DSB*, 4, 254–6.
Lemay, Richard, 'Gerard of Cremona', *DSB*, 15, 173–92.
Lindberg, David C. (ed.), *Science in the Middle Ages* (Chicago, 1978).
———, *The beginnings of European Science, the European scientific tradition in philosophical, religious and institutional context, 600 BC to AD 1450* (Chicago, 1992).
———, 'John Pecham', *DSB*, 10, 473–6.
———, 'Witelo', *DSB*, 14, 457–62.
Lindberg, David C. and Robert S. Westman (eds), *Reappraisal of the Scientific Revolution* (Cambridge, 1990).
Little, A. G. (ed.), *Roger Bacon Essays* (Oxford, 1914).
Lloyd, G. E. R., *Early Greek Science, Thales to Aristotle* (London, 1970).
———, *Greek Science After Aristotle* (London, 1973).
Lohne, J. A., 'Thomas Harriot', *DSB*, 6, 124–9.
Lucretius, *De Rerum Natura (On the Nature of the Universe)*, translated by W. H. D. Rouse (London and Cambridge, Massachusetts, 1937).
Lukowski, Jerzy and Hubert Zawodski, *A Concise History of Poland* (Cambridge, Massachusetts, and New York, 2006).
Lyons, Jonathan, *The House of Wisdom, How the Arabs Transformed Western Civilization* (London, 2009).
Machamer, Peter (ed.), *The Cambridge Companion to Galileo* (Cambridge, 1996).
Mahoney, Michael S, 'Descartes: Mathematics and Physics', *DSB*, 4, 55–61.
———, 'Pierre de Fermat', *DSB*, 4, 566–76.
Makdisi, George, *The Rise of Colleges: Institutions of Learning in Islam and the West* (Edinburgh, 1981).
———, *The Rise of Humanism in Classical Islam and the West* (Edinburgh, 1990).
Manuel, Frank E., *A Portrait of Isaac Newton* (Cambridge, Massachusetts, 1968).
Masson, Georgina, *Frederich II of Hohenstaufen, a life* (London, 1957).
Masotti, Arnaldi, 'Francesco Maurolico', *DSB*, 9, 190–4.
McCluskey, Stephen C., *Astronomies and Cultures in Early Medieval Europe* (Cambridge, Massachusetts, 1998).
McVaugh, Michael, 'Constantine the African', *DSB*, 3, 393–5.
Milton, John, *Paradise Regained*, edited by Merritt Y. Hughes (New York, *c.*1937).
———, *Paradise Lost*, ed. Edward Le Comte, New York, 1961.
Minio-Paluello, Lorenzo, 'Boethius', *DSB*, 2, 228–36.
———, 'James of Venice', *DSB*, 7, pp. 65–7.
———, 'Michael Scott', *DSB*, 9, 361–5.
———, 'William of Moerbeke', *DSB*, 9, 434–40.
———, 'Plato of Tivoli', *DSB*, 11, 31–3.

Molland, A. C., 'John of Dumbleton', *DSB*, 7, 116–17.
Monfasani, John, *Byzantine Scholars in Renaissance Italy: Cardinal Bessarion and other Emigres; Selected Essays* (Brookfield, Vermont, 1995).
———, *Greeks and Latins in Renaissance, Studies in Humanism and Philosophy in the 15th Century* (Brookfield, Vermont, c.2004).
Moody, Ernest A., 'Albert of Saxony', *DSB*, 1, 93–5.
———, 'Jean Buridan', *DSB*, 2, 603–8.
———, 'William of Ockham', *DSB*, 10, 223–30.
———, Moody, 'Galileo and Avempace: The Mechanics of the Leaning Tower Experiment', *Journal of the History of Ideas* 12 (1951), pp. 375–93, 375–422.
Murdoch, John, 'Thomas Bradwardine', *DSB*, 2, 390–7.
Murdoch, John and Edith Dudley Sylla, 'Walter Burley', *DSB*, 2, 608–12.
———, 'Richard Swineshead,' *DSB*, 13, 184–213.
Newton, Isaac, *Opticks, or a Treatise on the Reflections, Refractions, Inflections, and Colours of Light* (London, 1952).
———, *Principia (Mathematical Principles of Natural Philosophy)*, translated by I. Bernard Cohen and Anne Whitman (Berkeley, 1999).
North, John David, *The Norton History of Astronomy and Cosmology* (New York, 1995).
———, 'Richard of Wallingford', *DSB*, 11, 414–16.
O'Leary, De Lacy, *How Greek Science Passed to the Arabs* (London, 1949).
O'Neil, W. M., *Early Astronomy from Babylonia to Copernicus* (Sydney, 1986).
Osler, Margaret (ed.), *Rethinking the Scientific Revolution* (New York, 2000).
Pannekoek, Anton, *A History of Astronomy* (New York, 1961).
Pastor, Ludwig, *The History of the Popes: from the close of the Middle Ages, drawn from the secret archives of the Vatican and other sources*, 40 vols. (London, 1949).
Payne-Gaposchkin, Cecelia, *Introduction to Astronomy* (New York, 1954).
Permuda, Loris, 'Pietro d'Abano', *DSB*, 1, 4–5.
Pingree, David, 'Leo the Mathematician', *DSB*, 8, 190–2.
Plato, *Complete Works*, edited by John M. Cooper (Indianapolis, 1997).
Pliny the Elder, *Natural History*, 10 vols., translated by H. Rackham et al. (Cambridge, Mass., 1942–63).
Plutarch, *Plutarch's Moralia*, 15 vols., translated by Harold Cherniss and William C. Helmbold (Cambridge, Massachusetts, 1957).
Poulle, Emmanuel, 'John of Sicily', *DSB*, 7, 141–2.
———, 'John of Saxony', *DSB*, 7, 139–41.
———, 'William of St. Cloud', *DSB*, 14, 389–91.
Ptolemy, *Ptolemy's Almagest*, translated and annotated by G. J. Toomer, with a Foreword by Owen Gingerich (Princeton, 1998).
Ragep, F. Jamil, 'Copernicus and His Islamic Predecessors: Some Historical Remarks', in *Filozofski vestnik* XXV, #2 (2004), 125–42.
———, 'Ali Qushji and Regiomontanus: Eccentric Transformations and Copernican revolutions', in *Journal for the History of Astronomy* XXXVI (2005), 359–71.
Ragep, F. Jamil, and Sally P. Ragep with Steven Livesey (eds), *Tradition, Transmission, Transformation* (Leiden, 1996).

Rashdall, Hastings, *The Universities in Europe in the Middle Ages*, 3 vols. (London, 1936).
Repcheck, Jack, *Copernicus' Secret: How the Scientific Revolution Began* (London, 2009).
Rochot, Bernard, 'Pierre Gassendi', *DSB*, 5, 284–90.
Ronan, Colin A., *The Cambridge Illustrated History of the World's Science* (London, 1983).
———, 'Edmond Halley', *DSB*, 6, 67–74.
Rosen, Edward, *Three Copernican Treatises: The Commentariolus, The Letter Against Werner, and The Narratio Prima of Rheticus* (New York, 1st ed., 1939).
———, 'Was Copernicus a Neoplatonist?', in *Journal of the History of Ideas*, Vol. 44, No. 4 (Oct. 1983), 667–69.
———, *Copernicus and the Scientific Revolution*, Malabar, Florida, 1984.
———, *Copernicus and His Successors*, edited by Erna Hilfstein (London, 1995).
———, 'Federico Commandino,' *DSB*, 3, 363–5.
———, 'Nicholas Copernicus', *DSB*, 3, 401–11.
———, 'Michael Mastlin, *DSB*, 9, 167–71.
———, 'Andreas Osiander', *DSB*, 10, 245–6.
———, 'Johannes Regiomontanus', *DSB*, 11, 348–52.
Rosinska, Grazyna, 'Nasir al-Din al-Tusi and Ibn al-Shatir in Cracow?', in *Isis* 65, No. 2 (1974), 239–43.
Rubenstein, Richard E., *Aristotle's Children: How Christians, Muslims, and Jews Rediscovered Ancient Wisdom and Illuminated the Middle Ages* (New York, 2001).
Rudnicki, Jozef, *Nicolaus Copernicus* (London, 1943).
Runciman, Steven, *The Last Byzantine Renaissance* (Cambridge, 1970).
———, *Byzantium and the Renaissance* (Tucson, 1970).
Sabra, A. I., 'The Appropriation and Subsequent Naturalization of Greek Science in Medieval Islam: A Preliminary Statement', Ragep, F. Jamil, and Sally P. Ragep with Steven Livesey (eds) *Tradition, Transmission, Transformation*, pp. 3–27.
Saliba, George, *A History of Arabic Astronomy: Planetary Theories during the Golden Age of Islam* (New York and London, 1994).
———, *Islamic Science and the Making of the European Renaissance* (Cambridge, Massachusetts, 2007).
Sambursky, Samuel, *The Physical World of the Greeks* (London, 1956).
———, *Physics of the Stoics* (New York, 1959).
Samso, Julio, 'Levi ben Gerson', *DSB*, 8, 279–82.
Santillana, Giorgio de, *The Crime of Galileo* (Chicago, 1955).
Sarton, George, *Introduction to the History of Science*, 3 vols. in 5 parts (Baltimore, 1927–48).
———, *A History of Science*, 2 vols. (Cambridge, Massachusetts, 1952, 1959).
———, *The Appreciation of Ancient and Medieval Science During the Renaissance, 1459–1600* (Philadelphia, 1955).
Seneca, *Natural Questions*, translated by J. Clarke (London, 1910).
Schneidler, F., 'Johannes Schöner', *DSB*, 12, 199–200.
Scott, J. F., 'Christopher Wren', *DSB*, 14, 509–11.

Seneca, *Natural Questions*, translated by J. Clarke (London, 1910).
Settle, Thomas W., 'Giovanni Alfonso Borelli', *DSB*, 2, 306–14.
Shakespeare, William, *The Complete Works*, edited by Peter Alexander (New York, 1952).
Shank, Michael H., 'The Classical Scientific Tradition in Fifteenth-Century Vienna', in Ragep, F. Jamil, and Sally P. Ragep with Steven Livesey (eds), *Tradition, Transmission, Transformation*, pp. 115–36.
Shapin, Steven, *The Scientific Revolution* (Chicago, 1996).
Sharpe, William D., 'Isidore of Seville', *DSB*, 7, 28–30.
Singer, Charles, E. J. Holmyard, and A. R. Hall, *A History of Technology*, 2 vols. (Oxford, 1954–1984).
Spenser, Edmund, *The Faerie Queene*, 2 vols. (New York, 1927).
Stahl, William H., 'The Greek Heliocentric Theory and its Abandonment,' in *Transactions and Proceedings of the American Philological Society*, 76 (1945), 321–32.
———, 'Macrobius', *DSB*, 9, 1–2.
———, 'Martianus Capella', *DSB*, 9, 140–1.
Steel, Duncan, *Marking Time: The Epic Quest to Invent the Perfect Calendar* (New York, 2000).
Strabo, *The Geography*, 8 vols., translated by Horace Leonard Jones (Cambridge, Massachusetts, 1969).
Struik, D. J., 'Gerbert d'Aurillac', *DSB*, 5, 364–6.
Sullivan, J. W. N., *Isaac Newton, 1642–1727* (New York, 1928).
Swerdlow, N. M., 'Annals of Scientific Publishing: Johan Petreius's Letter to Rheticus,' *Isis*, Vol. 83, No. 2 (June 1992), pp. 270–4.
———, 'Copernicus, Nicolaus (1473–1543),' in *Encyclopedia of the Scientific Revolution from Copernicus to Newton*, ed. Wilbur Applebaum (London/New York, 2000, p. 165).
Swerdlow, N. M. and O. Neugebauer, *Mathematical Astronomy in Copernicus' De revolutionibus*, 2 vols. (New York, 1984).
Talbot, Charles H., 'Stephen of Antioch', *DSB*, 13, 38–9.
Taton, René, *History of Science*, 4 vols., translated by A. J. Pomerans (New York, 1964–6).
———, 'Blaise Pascal', *DSB*, 10, 330–42.
Thomas, Phillip Drennon, 'Alcuin of York', *DSB*, 1, 104–5.
———, 'Alfonso el Sabio', *DSB*, 1, 122.
———, 'Cassiodorus', *DSB*, 3, 109–10.
Thoren, Victor, E., 'John Flamsteed', *DSB*, 5, 22–6.
Thorndike, Lynn, *A History of Magic and Experimental Science*, 8 vols. (New York, 1923–56).
———, *Michael Scot* (London, 1966).
Thucydides, *History of the Peloponnesian War*, translated by Rex Warner (Harmondsworth, 1987).
Toomer, G. J., 'Campanus of Novara', *DSB*, 3, 23–9.

Van Helden, Albert, *Measuring the Universe: Cosmic Dimensions from Aristarchus to Halley* (Chicago, 1985).
Vescovini, Graziella, Federic, 'Francis of Meyronnes', *DSB*, 5, 115–17.
Vogel, Kurt, 'Byzantine Science', in *Cambridge Medieval History*, new edition, vol. 4, part 2, pp. 264–305.
———, 'Leonardo Fibonacci (Leonardo of Pisa)', *DSB*, 4, 604–13.
Wallace, William A., 'Saint Albertus Magnus', *DSB*, 1, 99–103.
———, 'Saint Thomas Aquinas', *DSB*, 1, 196–200.
———, 'Dietrich of Freiberg', *DSB*, 4, 92–5.
———, 'William of Auvergne', *DSB*, 14, 388–9.
———, 'Galileo's Pisan studies in science and philosophy', in Peter Machamer (ed.), *The Cambridge Companion to Galileo*, pp. 27–52.
Ward, Benedicta, *The Venerable Bede* (Harrisburg, Pennsylvania, 1990).
Webb, J. F. (translator), *The Age of Bede* (New York, 1988).
Westfall, Never at Rest, *A Biography of Isaac Newton* (Cambridge, Massachusetts, 1980).
Whitfield, Peter, *Landmarks in Western Science: From Prehistory to the Atomic Age* (London, 1991).
———, *From Byzantium to Italy: Greek Studies in the Italian Renaissance* (London, 1992).
Wilson, Curtis A., 'William Heytesbury', *DSB*, 6, 376–80.
Wilson, Nigel, From Byzantium to Italy: Greek Studies in the Italian Renaissance, London, 1992.
Yates, Frances A., 'Giordano Bruno', *DSB*, 2, 539–44.
Youschevitch, A. P., 'Isaac Newton', *DSB*, 10, 42–103.

INDEX

Abu Ma'shar 32
Adelard of Bath 32–3
Albrecht, Duke of Prussia 66, 89, 101, 102, 106, 133, 144, 147, 148, 185
Alcuin of York 30–1
Alexander I, King of Poland 66
Alexander VI Borgia, Pope 1, 54, 61
Alexandria 20–3
  Library and Museum 14, 21
Alfonso VI, King of Castile and León 33
Alfonso X, King of Castile and León 46
  Alfonsine Tables 46, 50, 51, 60, 90, 93, 187
Alhazen 33, 41, 46–7, 50
Allenstein 86–7
Anaxagoras 14–5
Anaximander 14, 17
Anaximenes 14, 17
Anthemius of Tralles 26–7
Apian, Peter *see* Apianus, Petrus
Apianus, Petrus 93, 94, 117, 122–4
Apollonius of Perge 22–4, 122, 194, 227, 231
Aquinas, Thomas, St 217
Aratus of Soli 23
Archimedes 16, 21, 22, 33, 34, 76, 122, 198, 206–8, 215, 231
Aristarchus of Samos 21, 22, 76, 77, 186, 188, 221
Aristotle 15, 17, 19, 20, 27–34, 38–43, 50, 56, 62, 76, 93, 94, 106, 125, 188–90, 191, 205–7, 217, 220, 230, 240
Athens 14–6, 19
  Academy 15, 16, 27
  Lyceum 15, 19
Augustine, St, Bishop of Hippo 28
Aurillac, Gerbert d' 31–2

Averroës 34, 38, 62, 145
Avicenna 33, 38, 41, 66

Bacon, Francis 226
Bacon, Roger 41, 46–7
Baghdad 27
  Bayt al-Hikma 27, 46
Banu Musa 33–4
Barrow, Isaac 231
al-Battani 50, 62
Bede, the Venerable 29–30
Beeckman, Isaac 225–6
Bellarmine, Robert, Cardinal 214–5, 218
Benedetti, Giovanni Battista 188
Bessarion, Johannes 54–7
Blundeville, Thomas 186
Boethius, Anicius Manlius Severinus 28–9
Bologna University *see* University
Boyer, Carl B. 57, 150
Boyle, Robert 230–1
Bradwardine, Thomas 42
Brahe, Tycho 74, 75, 189, 196–7, 205, 218–9, 225, 243, 245
Braunsberg (Braniewo) 8–9, 12, 86
Brudzewski, Albert 50, 79
Bruno, Giordano 188
Buonaccorsi, Filippo, 11
Buonamici, Francesco 206, 207
Buridan, John 42–3

Callipus of Cyzicus 17
Cambridge University *see* University
Camerarius, Joachim 117, 124, 125, 155–6
Cardano, Geronimo 123–4
Casimir I, King of Poland 37
Casimir III, King of Poland 4, 48, 49

Casimir IV, King of Poland 5, 8–11
Celtes, Conradus 10–1
Charlemagne, Holy Roman
  Emperor 30–1
Christian IV, King of Denmark 192
Clavius, Christopher 214
Clement VII, Pope 106
Cologne University *see* University
Columba, St 29
Columbus, Christopher 51, 57
Constantinople 13–4, 26–7, 53–5
Copenhagen 189–90
Copernicus, Andreas 7, 8, 12, 37, 49,
  53–4, 60–3, 65, 108
Copernicus, Barbara 7, 8
Copernicus, Katherina 7, 8
Copernicus, Nicolaus
  administrative duties as canon, 85–7
  Allenstein Bread Tarrif 99, 100
  astronomical instruments 70, 71,
    248
  astronomical observations 59–61,
    68, 70, 71
  childhood 2, 6–8
  collaboration with Rheticus, 131–45,
    147–62
  Commentariolus 73–4, 75–84
  De revolutionibus 59, 60, 62, 63, 68,
    72, 73, 76, 78, 81–4, 93–6, 105,
    120, 121, 124, 147–62, 163–83,
    245, 246
  duties as diplomat 65, 66
  education of 1, 7, 8, 12, 37, 48–51,
    57–61
  Epistles of Theophylactus Simocatta 60,
    67
  final illness and death 159–60
  Greek predecessors 21–4, 31, 48, 51,
    76, 77, 186, 188
  Islamic predecessors 33, 50, 51, 57,
    172, 173, 179
  Letter Against Werner 85–97
  Narratio prima 58, 60, 62, 115, 118,
    126, 129, 131–45
  other Latin translations 67–8
  relationship with Anna
    Schilling 103–4, 110–3, 126, 161
  in Rome 60–1
  self-portrait of ii, 247
  tomb of 160–1, 247–49
  work as a cartographer 101, 115,
    147, 148
  work as a physician 65–7, 100, 104,
    147
  work on coinage 88, 89
Cosimo II de' Medici 209, 214, 217
Ctesibus of Alexandria 23

Danielson, Dennis 151, 246
Dantiscus, Johannes, Bishop 107–13,
  126, 134, 159–61
Danzig (Gdansk) 2, 3, 6, 7, 68, 102,
  107, 131, 144
Dasypodius, Conrad 247
Dee, John 187
Democritus 15, 20, 28
Descartes, René 226–31, 235, 239, 243
Dietrich of Freiburg 47, 48, 228
Digges, Thomas 187–8
Diophantus 26, 227
Dioscorides 26, 66
Dlugosz, Jan 10–1
Donne, John 201
Donner, George 159, 161
Drake, Stillman 215, 217

Elbing (Elblag) 65, 68, 69, 85, 86, 89,
  102–4
Empedocles 14, 17
Epicurus 20
Erasmus, Desiderius 125–6
Eratosthenes 21
Euclid 21, 22, 26, 56, 194, 206
  Elements 29, 33, 50, 122
Eudoxus 16–7
Evans, James 119

Fabricus, Johannes 216
Al-Farabi 33
Al-Farghani 33, 50
al-Farisi, Kamal al-Din 48
Feild, John 187
Ferber, Maurice, Bishop 89, 100–6
Fermat, Pierre 227
Ferrara University *see* University
Florence 205–6, 214–5, 222

INDEX

Foucault, Jean-Bernard-Léon 248
Frauenburg (Fromberk) 53, 61, 68–71, 84, 86, 87, 97, 100–2, 129, 135, 148, 149, 160, 161, 246
Frederick II, King of Denmark 190, 192
Frisius, Gemma 134

Galen 25, 26, 33, 34, 38
Galilei, Galileo 44, 195, 200–1, 205–24, 231
Gassendi, Pierre ii, 2, 202, 225, 231
Gasser, Achilles Pirmin 116, 131, 134–5, 155
Geber 33, 50, 123–4
Gerard of Cremona 33, 34, 122
Giese, Tiedemann 65, 85, 99–101, 108, 112, 125–29, 133, 135, 144–7, 159–61, 164, 247
Gilbert, William 188
Gingerich, Owen 76, 77, 121, 126, 185–6, 246
Grassi, Horatio, Father 218–9
Gregory IX, Pope 38
Grosseteste, Robert 39–42, 46–7
Guericke, Otto von 230

Halley, Edmond 236–7
Hanseatic League 6–8
Harriot, Thomas 199, 201, 216
Hartman, Georg 149–50
Harun al-Rashid, Caliph 27
Heilsburg (Lidzbark Warminski) 12, 53, 65, 68, 86, 106, 109
Heraclides, Ponticus 19
Heraclitus 14
Hero of Alexandria 23, 34, 47
Heytesbury, William 42
Hipparchus of Nicaea 23–4, 31, 48, 79, 90
Hippocrates 15, 38
Hooke, Robert 230, 234–7, 243
Hunayn ibn Ishaq 41
Huygens, Christiaan 228–34

Ibn al-Haytham *see* Alhazen
Ibn Rushd *see* Averroës
Ibn al-Shatir 172–3

Ibn Sina *see* Avicenna
Isidorus of Miletus 26–7

Jabir ibn Aflah *see* Geber
Jabir ibn Hayyan 33
Jadwiga, Queen of Poland 4–5
James I, King of England 201
John I Albert, King of Poland 66
John of Holywood *see* Sacrobosco, Johannes de
John of Werden 126
Julius II, Pope 70, 71
Justinian I, Emperor 14, 26

Kepler, Johannes 181, 193–203, 208, 239–40
al-Khwarizmi 32–3
al-Kindi 33, 50
Königsberg 6, 56, 147
Koppernigk, Barbara *see* Copernicus, Barbara
Koppernigk, Niklas *see* Copernicus, Nicolaus
Krakow 2–12
Krakow University *see* University
Kulm (Chelmno) 3, 65, 108–13

Leibnitz, Gottfried Wilhelm 242
Leipzig University *see* University
Leo X, Pope 71–2
Leonardo da Vinci 54, 59
Lossainen (Luzjanski), Fabian von, Bishop 69–70, 100
Luther, Martin 101–2, 153–5

Maestlin, Michael 158
al-Mahdi, Caliph 27
al-Ma'mun, Caliph 27
al-Mansur, Caliph 27
Maragha School 172–73
Marienburg (Malbork) 65
Masha'allah 33, 50
Matthew of Miechow 49, 73–4
Matthias Corvinus, King of Hungary 9, 56
Maximilian I, Emperor 11, 71, 104, 108
Melanchthon, Phillipp 116–18, 124–5
Mersenne, Marin 226

Napier, John 151, 202
Nemorarius, Jordanus 41
Neugebauer, Otto 68, 148, 150, 172–5, 178–9, 181–82, 246; *see also* Swerdlow, Noel
Newton, Isaac 231–43
Nicholas of Cusa, Cardinal 48
Niederhoff, Leonard 111, 127, 161
Novara, Dominico Maria de 58–60
Nuremberg 101, 117–9, 124, 131, 150, 155–6

Oldenburg, Henry 233–35
Oresme, Nicole 42–5
Osiander, Andreas 124, 131–32, 156–7
Oxford University *see* University

Padua University *see* University
Pannekoek, Anton 120, 122
Paris University *see* University
Parmenides 14
Pascal, Blaise 230
Paul II, Pope 1
Paul III, Pope 155, 163, 164, 165
Paul V, Pope 218
Paul of Middleburg, Bishop 72–3, 95
Peckau, Johann 2
Peregrinus, Petrus 45
Pericles 15
Petreius, Johannes 118, 124–5, 131, 133, 150, 155–9
Peurbach, Georg 48, 50, 51, 56–7, 118–9
Philo of Byzantium 23
Philoponus, John 27
Pisa 205–8
Pitiscus, Bartholomaeus 150–1
Plato 15, 16, 19
Plethon, George Gemistos 54
Prague 192–3
Ptolemy (Claudius Ptolemaaus) 23–6, 33–4, 40, 41, 46, 48, 50, 51, 55–62, 70, 71, 75, 79–81, 90–3, 97, 122–4, 135, 145, 149, 150, 153, 163, 168–74, 178–83, 185–9, 195, 200, 205, 226
Ptolemy I, King of Egypt 21

Ptolemy II, King of Egypt 21
Ptolemy III, King of Egypt 21–2
Ptolemy IV, King of Egypt 22
Pythagoras, Pythagoreans 14–5, 19, 164, 167

Qushji, Ali 57
Qusta ibn Luqa 33
Qutb al-Din al-Shirazi 172

al-Razi 33, 50
Recorde, Robert 186
Regiomontanus, Johannes 48, 50–1, 56–7, 118–20, 150
Reinhold, Erasmus 117, 149, 185–6
Rheticus, Georg Joachim 58–62, 73, 115–29, 131–45, 147–61, 185
Roemer, Olaus 241–2
Rome 1, 10–1, 25, 29–32, 53–5, 60–1, 35, 70–2, 101, 103, 113, 155, 188, 206, 214–9, 222
Rosen, Edward 58, 67, 74, 75–8, 91–6, 145, 155, 245–6
Rudolph II, Emperor 192, 197, 200
Rudolphine Tables 198, 200, 202

Sacrobosco, Johannes de 41
Scheiner, Christopher, Father 215–6
Schilling, Anna 103–4, 110–3, 126, 161
Schönberg, Nicolaus, Cardinal 105–6, 155–6, 164
Schöner, Johannes 117–29, 131–6
Scultetus, Alexander 100, 101, 108–13, 115, 127, 148
Sigismund I, King of Poland 66, 69, 70, 85, 86, 88, 89, 102, 107, 108, 113
Simplicius 16, 19, 26, 27, 39
Socrates 15
Soto, Domingo de 209
Stevin, Simon 206–7
Stimmer, Tobias ii, 247
Swerdlow, Noel 68, 148, 150, 172–5, 178–9, 181–82, 246; *see also* Neugebauer, Otto
Sylvester II, Pope *see* Aurillac, Gerbert, d'

Teschner, Philipp 8, 12
Thabit ibn Qurra 33, 46, 50, 72, 90, 93
Thales 14, 17
Theodosius of Bithynia 24
Theon of Alexandria 26, 46, 122
Theophrastus 19–20
Thorn (Torun) 2–12, 69, 86, 102, 246, 247
Toledo 33
Torricelli, Evangelista 230
Trapezuntios, George 55–6
al-Tusi, Nasir al-Din 50, 172

University
  Bologna 1, 8, 38, 53, 54, 58
  Cambridge 38
  Cologne 8, 10
  Ferrara 1, 63
  Leipzig 156, 185
  Krakow 37–52
  Oxford 38, 39, 41, 42
  Padua 1, 38, 62, 63
  Paris 36, 38–41, 47
  Vienna 50, 56
  Wittenberg 115–6, 125, 147–9, 156, 185
Urban V, Pope 48–9

Urban VIII, Pope 218–22
al-'Urdi, Mu'ayyad al-Din 172, 179

Vienna University *see* University
Vögelin, Georg 134–5

Walther, Bernhard 119–22
Wapowski, Bernard 49, 89–91, 105
Warmia 6, 9, 12, 65, 66, 69, 70, 85–9, 101–9, 125–9, 159, 161
Watzenrode, Lucas, the Elder 2, 6
Watzenrode, Lucas, the Younger 2, 3, 8, 9, 12, 53, 60, 65, 100, 160
Werner, Johannes 89–96
William of Moerbeke 34, 47
William of St Cloud 46
Witelo 46–7
Wittenberg, 115–6, 124–5, 148–9, 153–5
Wittenberg University *see* University
Wladyslaw II Jagiello, King of Poland 5, 49

Xenophanes 14

Zell, Heinrich 101, 115, 131, 147–8
Zeno of Citium 20